Collection Development in the Digital Age

Collection Development in the Digital Age

Editor

Dominika Dechiel

Collection Development in the Digital Age

Edited by **Dominika Dechiel**

ISBN: 978-1-68117-254-5
Library of Congress Control Number: 2016934797

© 2017 by
SCITUS Academics LLC,
www.scitusacademics.com
Box No. 4766, 616 Corporate Way,
Suite 2, Valley Cottage,
NY 10989

Preface

The term "collection development" refers to the process of systematically building library collections to serve study, teaching, research, recreational, and other needs of library users. The process includes selection and deselection of current and retrospective materials, the planning of strategies for continuing acquisition, and evaluation of collections to determine how well they serve user needs. Overall, collection development encompasses many library operations ranging from the selection of individual titles for purchase to the withdrawal of expendable materials. An impressive number of publications on collection development have been produced within the last twenty years.The collection development policy is intended to provide guidance, within budgetary and space limitations, for the selection and evaluation of materials which anticipate and meet the needs of the Pasadena community. It directly relates the collection to the library's mission statement, and defines the scope and standards of the various collections. As the community changes, the library will need to reassess and adapt its collections to reflect new and differing areas of interest and concern. The collection development policy will be periodically evaluated and revised as necessary to provide guidance for implementing changes in the collection. Before entering into a discussion of the policy framework, it is important to recognize the issues currently affecting the collection development planning process. The special library is usually created for the specific purpose of providing accurate and current information for a particular set of patrons and, therefore, must contain materials considered to be of quality and be able to access them in sometimes demanding circumstances. Inaccurate information in a special library may cause destruction for the supporting institution. The second issue involves concern over the information explosion. The amount of research and number of researchers is increasing, and the amount of published material is increasing, thereby challenging libraries in their desire to provide access to this increased wealth of information. Meanwhile, libraries have also been trying to contend with the rising cost of the materials in conjunction with the decreasing space in their physical facilities. Additionally, the impact of technology has created problems in accessing information. This book entitled Collection Development in the Digital Age explores evolving roles and ideas in collection development in the digital age.

Table of Contents

CHAPTER 1

Library Promotion Practices and Marketing of Library Services: A Role of Library Professionals

S. K. Patil, Pranita Pradhan*

Central Library, Symbiosis International University (SIU), Lavale, Pune - 412 115, Maharashtra State, India

ABSTRACT

Marketing of products in industries is very much essential to increase sale and consequently to gain the profit, however in academic environment like university and colleges promotion and marketing of library and information sector is required to aware the library services. This paper enumerates the concept of library promotion and marketing of library services. Academic libraries are having all type of collections. To promote the collection in use it is necessary to market library products. This paper suggests practical solutions, ways and means of marketing the library services.

KEYWORDS

Library promotions; marketing of library services; Role of library professionals; Library products and services

1. INTRODUCTION

Libraries are considered as treasures of knowledge. It is also known as storehouse of knowledge. It is true that all libraries in the world are full of reading material which consist books, journals, films, images, manuscripts, Audio visual materials etc. which has knowledge, recorded by peoples, eminent writers and eminent personalities. Preservation of this knowledge is a prime task of all libraries and Information Centers; however it is a need to market these resources which possess by the libraries. Now a day with the help of Information technology many libraries and information centers has their own websites on which all kind of material is accessible with its bibliographic details and mechanism to promotion and marketing of services. Marketing concepts in Library Services

2. MARKETING CONCEPTS IN LIBRARY SERVICES

Philip Kotlar, Marketing Guru has define marketing "as social and Managerial process by which individuals and groups obtain what they : need and want through creating, offering and exchanging product of value and others ". In view of the above definition, library activities are a team work or the efforts of group working I library. To attract more and more users to the library, Library staff needs to extend promotion and cooperation to users and marketing their services. The basic purpose behind promotion is to educate the students, faculty members in how to use the library and its resources and also to upkeep their knowledge by providing information appended in various sources available in the library. Like Companies promotion and marketing concept, library promotion and marketing services are different. The primary purpose of marketing of company products is to increase sales and ultimately to gain the more profit from it. The libraries are non-profit organizations; It is a social organization and service centre.

3. NEED OF LIBRARY MARKETING OF LIBRARY SERVICES

To promote the use of available reading material in the library and create awareness among the users.

To optimize the use of information within limited resources and manpower.

Limited Budget for library needs to market services and generate funds for library

To improve the image of the library.

Due to information explosion, readers require precise and correct information for their research and study.

Unless and until what is available in the library, how it would be accessible and disseminated to the users of the library. The Users do not aware the resources available in the libraries. Now days libraries are investing huge amount to purchase reading material and subscription to periodicals and online databases to fulfil the needs of their students, faculty members and research scholars. Investing such amount for the resources, the usage of these resources should also increase. Libraries should think and work out the cost benefit analysis of this investment. Therefore it is quite necessary to literate people about the services providing by libraries and promote its use.

There are several ways to promote the usage and marketing of services

Organization of Information Literacy programme on regular basis at various level.

Organization of workshops/ training programs about awareness of resources available in the libraries and Information centres.

Organization of Training programs to library staff with modern technologies and expertise people.

Attract the people by organizing book exhibitions of new books with the help of vendors or the material available in the library should be displayed at prominent place.

4. OBJECTIVES OF LIBRARY PROMOTIONS AND MARKETING OF LIBRARY SERVICES

- The basic objective behind library promotion and marketing of library services and products are to achieve high level customer's satisfaction, ensure the survival of their respective institutions and enhance the perceived value of the services.
- Libraries can promote the use of the services and products but cannot make profit out of it because library is a social institution.

5. CUSTOMER SEGMENTATIONS

The main segmentations of the library marketing are their users, i.e. students, faculty members, staff, Research scholars, General users of the library, International students studying in various courses etc. It is difficult to work out single marketing strategy for all the category of segments. However it will be worthwhile to formulate different strategies for individual group of segments.

6. DR. RANGANATHAN CONCEPT OF MARKETING OF LIBRARY SERVICES THROUGH FIVE LAWS OF LIBRARY SCIENCE

Dr. S.R. Ranganathan, father of library and Information Science, philosopher, mathematician devised five laws of library science which promotes the usage of resources.

Books are for use :

This law itself promotes that each book available in the library is for use. If the books kept in lock and key without providing it to the readers, then it is a dead investment of the organization. Hence books should be freely available to each user coming to the library. Library staff should take initiative to attract their users to read more and more books. But while giving more books to the users, they should see how it come back for further use to other readers. It is a skill of the library staff , how the attract the students/customers towards the library. Books are for user promotes the use of book. Staff should tell the users what are the good books available, it should be placed on the prominent place , may be entrance of the library where every user come and see.

Every Reader his book :

Here Reader of the library is main factor/customer, every staff must see how library users are satisfied with the services by the libraries. We must focus to the needs of such readers and their satisfaction. It is true that library cannot satisfy each and every customer, however efforts should be taken by the staff to satisfy users maximum. **Every book its reader:**

According to this concept/law Dr. Ranganathan expressed that every book which is purchased by library must get reader. Here we must see what the needs of the readers are. Dr. Ranganathan expect that find a reader for every book. To find a reader for every book library staff should conduct such studies/surveys to obtain the needs of their readers. Once you identify their needs then you can promote the resources and increase the use of it. Here attitude of the library staff is more important. They should be always positive to solve the problems of the users with smiling face.

Save the time of a reader :

The users time is very precious, to save the time of users in the library, staff should organise the information in such a way to fine it promptly. Reader should not waste their time in searching the information, searching the books and journals in the library. Of course arrangement of library material is a scientific method which is taught to the library staff.

Library is a growing organism :

Here more emphasis is on evolutionary growth of the library. However librarian must see how library collection grows with qualitatively not in quantity. Now a days e-books are more popular and easy to access. Hence vendors are promoting and marketing their products in package or pick and choose model where choice of selection of required books are with faculty members.

7. SEVEN PRINCIPLES OF MARKETING IN TERMS OF LIBRARY AND INFORMATION SCIENCE

There are seven principles of marketing; these principles are also applicable to Library and Information services. These principles are Product, Price, Place, Promotion, Participants, Physical evidence and Process. These principles are described in brief.

Product : Academic Libraries are providing good number of services to their users through various ways e.g. List of Additions (Whatever material added to the library it is communicated to users either printed list of through email). Now OPAC is accessible to all users through internet.

Current Content Service, SDI service, Web based services etc. These services are the product of libraries. Hence library professionals have to promote and market their products among the users.

Price : Price factor is important from the point of view of the budget of the Institution. Without budgetary provision, no library cannot purchase reading material. No library cannot self-sufficient By considering the academic out of the institution, faculty improvement, students growth institution has to make the provision in their budget. Not even for the reading material but for providing qualitative services, it require financial provision. Internet connectivity should be provided to students with no extra cost.

Place: Services are intangible; they cannot be normally stored, transported or inventoried. Services production cannot be separated from selling. In case of library services, personalized services like SDI are provided to the selected users by collecting what are the areas in which they require information. Library instructions shall be conducted either in the library or computer laboratory to provide hands on training to all segments.

Promotion: Promotion is another important phenomenon in marketing. It requires mechanism by which target groups are informed about the resources available in library and Information Centre. Promotion of what we have in the library. Users may not be aware or familiar with the library system. Hence it is essential that every year new students are joined with the library and at the beginning of the academic year, they must be provided awareness programmes. Word of mouth campaign is the best mechanism for promoting the user of library services. The primary promotion tool is library instructions in the form of workshop, seminars, lectures etc.

Participants : The success of any programmes is depend on the feedback of the participants hence participants involved in promoting and marketing of the library services provided by library professionals, their feedback will help to get the lacunas in the system and it will help to improve the services and library system.

Physical Evidence : According to Shostack " a physical object is self-defining ; a service is not " Hence in educational sector the marketing task is " defining for the services what it cannot define for itself " Evidence for the service can be both peripheral and essential. Physical evidence can support the marketing programmes by providing adequate service to the library users. It can make the service tangible.

Process: process is related to the process management, it consists of process planning , control, operation planning, facilities to be available with users, scheduling, quality of services etc.

8. WAYS AND MEANS OF PROMOTION AND MARKETING OF INFORMATION SERVICES AND PRODUCTS

To develop the products e.g. creation of databases, Union Catalogue, etc.

To give wide publicity among the users, institutions affiliated to universities, list of additions are to be publishing in local newspapers etc.

To establish good public relation with user community.

To conduct surveys of the users to know their needs, requirements from the library and improve the services and research products as per the requirements.

To provide carrier guidance and counseling service by the library to users.

9. PROFESSIONALS SKILLS REQUIRED FOR MARKETING LIBRARY/INFORMATION SERVICES.

Perception of user's needs, knowledge of conducting users surveys etc.

Ability to obtain feedback from users and skill to analyze the feedback.

10. What Library Professionals should do for Marketing of library and Information Services? (Practical solutions)

9

Most important quality of taking right decision at right time.

He must have technical knowledge such as use of internet, web page design, and product design and presentation skills.

He should have knowledge of various marketing strategies

10. WHAT LIBRARY PROFESSIONALS SHOULD DO FOR MARKETING OF LIBRARY AND INFORMATION SERVICES? (PRACTICAL SOLUTIONS)

Create awareness among the user and library staff.

Create awareness to offer services and products and expertise.

To know the users need and find out why they require information so that their Purpose can be identified.

Find out the users who do not use the library so that we can concentrate to such user and turn them to use library.

Use of mass media i.e. Radio, Television, AV Cassette, Video Programmes prepared for users .

Provide specialized services to special group of users e.g. Senior Citizen , blind users, Physically handicapped community users.

Organize talks of experts, seminars, debate, cultural programs etc. to attract users.

11. BEST INFRASTRUCTURAL FACILITIES AND BEST PRACTICES TO BE FOLLOWED BY LIBRARIES

Creation of Compact storage Section to accommodate more collection.

Student internship programmes. Students should be given opportunity to work in the library.

Organization should implement Earn while Learn Scheme for students so that are aware with the library collection, facilities, services etc.

Reading Hall facility should be expended 24 hours.

Orientation/ user education

Frequent meetings with library Staff,

Good bandwidth Internet access facility be made available with Wi-Fi Connectivity to all users on the campus.

Online access to E-resources should be made available through library Gateway portal.

Creation of digital resources

12. IMPLICATIONS FOR DEVELOPING MARKETING STRATEGIES

The main objective of the marketing of the library services to promote resources available in the library and Information Centre, another aim is to help in developing the career of the students. The library literature provides substantial amount of cases where libraries maintain close relationship with career services e.g. University of Pune library provide reading facilities for students who are studying for competitive examinations. Hollister (2005) describe how he was able to cultivate continuous relationship between the library and the University career services office at the University at Buffalo. Glynn & Wu (2003) clarified that library instructions has been strongly advocated as a tool to build liaison relationship not only with faculty but also with non-academic units.

13. CHALLENGES OF MARKETING IN LIBRARY AND INFORMATION SERVICES

To accept the challenges, libraries have to conduct surveys of the users and on the basis of these surveys they have to analysed the needs of their users and accordingly acquire the reading material, products, online databases, e- books etc. To make the collection is heavy user, libraries have to convert their print collection into electronic media and made it available to their users by considering the copy right issues. There is lack of financial budgetary provisions in the libraries. Every year budget is not increased proportionately. Continuous training programs are to be organised for students and staff of the library.

14. PLANNING FOR PROMOTION AND MARKETING OF LIBRARY SERVICES.

Every year new batches of students are admitted in the Institution, they are not well versed with the existing system of the library. Therefore it is a responsibility of the libraries to work out the planning of promotion and marketing of library services at the beginning of the academic year and it should be strictly followed according to plan. Each batch of student have to take a round in the library and library staff should brief them about what library possess, how to use, what are the facilities available.

15. CONCLUSION

Librarians and library staff are trying to find out the appropriate ways to respond the contemporary requirements of the students and faculty to fulfil their goals, meet the needs of existing and potential users.

Librarians are building the image and value of the libraries. The limited budget provision and the advent of new technology and its application in libraries have opened new vistas for marketing of library resources, products and services. In case libraries and library professionals fails to catchhold the opportunities, the opportunities will be grabbed by Commercial vendors and technology specialists. NAAC while accessing the institution or University more weightage has been given to library services. Now a days many libraries are preparing for marketing of library services and created a record for each event for accreditation. Therefore academic libraries, public libraries should plan their marketing and promoting the resources needs careful planning and policy and its execution on regular basis. The provisions of services should meet the requirements of current trends, respecting and identified and changing demands of more challenging users. Even in the curriculum of the Master's program of Library and Information Science discipline, separate papers on marketing and promotion of library services should be thoroughly taught. Library staff must realize the importance and understanding of marketing of services process, its theory and applications in practice such as Web.2.0 to promote activities and library mission to build a library brand name. New technologies such as YouTube, blogs, wikis attract the young users, provide an opportunity for libraries to revamp services.

REFERENCES

1. Kotler, Philip ; marketing of non-profit organizations. Ed.2 New Delhi : Printice-Hall of India, 1985
2. Coote, H. and Batchelor, B.: How to market your library services effectively. Ed.2 London : Aslib, 1997 p.19

3. Seetharama, S : "Guidelines for marketing for information services and products " In Marketing of library and information services in India.

4. Calcutta : IASLIC. 1988 pp 1-16.

5. Dragon, A.C : " Marketing the library ". Wilson library bulletin 53(7), 1979 pp.498-502

6. Hollister, C. (2005) . Bringing information literacy to career services, Reference Services Review, 33 (1), pp104-111.-

7. Glynn, T. & Wu, C. (2003) . New roles and opportunities for academic library liaisons : a survey and recommendations, Reference Service Review ,31 (2), pp122-128.

8. Zachert, M.J. and Williams, R.V. : " Marketing measures for information services ". Special Libraries, 17, 1986, pp 61-70

CHAPTER 2

Exploring utility function in utility management: an evaluating method of library preservation

Bin Yan[1*], Feng Shi[2] and Rui-qiang Yu[3]

[1]*Library, Nanjing University of Posts and Telecommunications, Nanjing, Jiangsu Province 210003 P. R. China*
[2]*Education College, Shandong Normal University, Jinan, Shandong Province 250014 P. R. China*
[3]*Economy and Management College, Nanjing University of Posts and Telecommunications, Nanjing, Jiangsu Province 210046 P. R. China*
Bin Yan,

ABSTRACT

In order to seek a new method of book evaluation and realize book resources sharing among the regional university libraries, we think that library should collect books of the high utility value in the case of limited funds. We proposed a changing Bellman equation as a utility function and used the explicit functions of the book usage factor and the book usage half-life derived from the utility function as an evaluating method of the collecting books. The results from empirical data given some conclusions such as the varieties diversity, the collected risk, the tendencies of reading varieties, species continuity and so on and a librarian can use the utility management to supplement the collections management.

KEYWORDS

Book evaluation Measure model Utility management Bellman equation Library preservation

1. INTRODUCTION

The relationship between supply and demand is undefined in current books from existing editing and publishing. The financing and product prices are becoming competing target among the resources consortia, which result in the cycle and flooding of the phenomena in home market, such as a single kind, vulgar content, lack of language and poor performance etc. Not only are these not able to meet the needs of the multi-level crowds, multi-regional custom, multi-species innovation, multi-fields academic inquiry as the scale, but these also don't reflect adequately the cultural prosperity and technology in the digital age. However, throughout the book supply chain, it is common that publishing group is to pursue profit maximization and authors are to compile the non-standard with the same materials from different publishers. The single monograph published with depth academic value and low sales is required of high publishing fee self-provided and underwriting these books, so that the domestic book market is hard to find ingenious academic works.

In recent years, library as the main preservation place of literature resources has increased in the number of annual collection with investment increase. But the number of collection of original academic books in Chinese has become less and less. The translation, reprint, repeatedly printing books and leisure books have been full of the library. The developing cultural resources of the network have led a large number of printing books to be in idle. There has been a decline because a lot of readers borrow printing books from the library. The electronic publications have become the most widespread form of publication. A largely open access to electronic journals and free e-book chapters make many readers often use electronic literature resources. That also contributes to the increase of the amount of collection and utilization for electronic literature resources.

Print books are also preferred by readers of students, researchers, children and elders. But as a result of the lack of scientific analogy between print and electronic of books, practical data can not be getting to proof the relationship between reading books and publishing books like academic journals. The results of book evaluation are one-sidedness from "ranking list" of the sales volume of books and are not believable the amount of loan or return books from the traditional methods (Ferne 1993) and statistic formulas of reading books (Yan 2009a, 2009b). Another book evaluation is an uncertainty of artificial interfere factors to manufacturing requirement by an author signing book to sell or a social celebrity reviewing book. Other date of book evaluation is unpredictable immediately with books rewarded by all-leveled governments and society organizations. There are other causes of incomplete data collected from selling and printing books, and little representation because the amount of reading and citation get less. These are unable to form an evaluation standard of all kinds of books.

K. H. Marx (1963) said long ago that every useful article "can be effective in a variety of different ways", so that, "a variety of ways to use article has been found"; "a utility of the article is endowed it with use value"; "use value is only realized in the usage or consumption". Utility, as one of the most commonly used concepts of economics, is a measure of satisfaction which the consumer get when she/he meets oneself demands and desires through consumption or enjoyment, also the capability that goods can meet person's desires. The utility function represents the function relationship between the numbers of the utility obtained by consumers and the p for consumer goods in consumption. Therefore, the utility in the utility value mainly means the desire of the consumer's subjective feeling for goods, which the inner strength to meet human needs (Say 1982). More precisely, utility refers to subjective enjoyment or usefulness a person got from consuming of an article or service (Paul and William 1996). The introduction of the utility to manage as utility management can analyse the effective use for the usage property of articles. Their utility contains the composition of the effective use outside the preferences degree of consumers for goods consumed.

The theory is not yet ripe from utility given out a management. The earlier utility management (Zhang and Wang 2005) should have both utility assessment and utility control to guide education as the conscious of the utility management (Xiao 2003), to be an aim of the management by the utility maximization (Li et al. 1998). Utility Management Automation System is a sort of supervisory control and data acquisition system for the electric utility system (Fereidunian et al. 2009). The utility from the economics were applied broadly in finance and insurance about risk and consume such as Markowitz combining model (Feng 2010). Using utility in library was applied to utility function about purchasing books (Lin 2005) and utility modeling of book usage (Zhang 2009) and utility statistics of books (Yan 2009a) and the utility management for the electronic resources (Yan and Xu 2010). The using utility after customer obtaining goods was not involved but the utility were only as the contents of the management. All we mentioned above conveyed a motion of management utility rather than the utility management that had not the utility as the main purpose of the management and also had not established the relation between utility function and applied analysis. It is an issue of the utility management in the article management. We dedicate to using utility management to the possession of articles in the utility value, as the premise of the management with the effective usage of empirical data to analyze the usefulness of continuing degree of these, and to mine the potential usefulness, to make full use of these, the maximum of use impact and management effectiveness (Yan and Wu 2011).

Collecting books in library after being rid of commercial property can express one's using utility. The utility function is a mathematical tool to express book using utility. We hope to seek a new mathematical measure model as utility function that is one part of the utility management and explore to apply the model to book evaluation. The librarian can use the utility management to supplement the current collection management.

2. LITERATURE REVIEW

Utility, we all have known, was proposed by Daniel Bernoulli when he explained St. Petersburg paradox in 1738. Its aim of the theory was to challenge a standard of decision-making with money desire value. In 19th century, early economists, such as William Stanley Jevons, Leon Walra and Alfred Marshall, believed that utility could be measured like the height and weight of people, while John Hicks insisted that under the assumption of ordinal number utility, the utility number was only expressed the order of preference, not the absolute number of utility. After the marginal utility proposed by Jevons, Menger and Walras in early 1970s, it was developed into a value theory by Böhm-Bawerk and Wieser. It is the feature to explain the process of value formation through the subjective psychology. Also, they thought that the value of goods was a kind of feeling and appraisal of people about the utility, which was decided by the utility and rare of goods. Total utility is the total satisfaction that consumers got from the certain consumption of goods and service over a certain period.

3. UTILITY FUNCTION

If the utility is the satisfactory degree of goods or services for consumer, then utility function means the function relationship in number between utility and the preference of goods. Let the utility function $U(x)$ as satisfactory model for x in the choice set for each to specify degree of preference. Generally $U(x)$ is between 0 and 1 (Farquhar 1984), which measures satisfactory degree the consumers feel by the consumption of specified combining goods. The utility function, we also have known, has get many mathematic forms such as the power utility function (Jana and Pavol 2010), the exponential utility function (Bhuvaneswari and Seethalakshmi 2011), Epstein-Zin utility function (Tan et al. 2006) etc., or is a cardinal utility function and a ordinal utility function. And the utility function has also been applied to many fields such as the information utility function, the multi-attribute utility function, the rent-seeking utility function, the traffic utility function, the objective utility function, the cost utility function, the HARA (hyperbolic absolute risk aversion, absolute risk aversion) utility function, the PD (partial distribution, partial tail distribution) utility function, the PRA (power risk aversion) utility function, the Logit (logistic) utility function and so on. These functions were defined from implicit function hypothesized to explicit function applied.

Assuming that the letter n is types of goods consumed and symbol x is consumption vector, and the utility function is the $U(x) = F(x_1, x_2, \cdots, x_n)$. The function is not an empty set, owing to goods consumed actually, and the preference is monotonic relationship. In the real domain of Euclidean space, the $U : R^n_+ \to R$, the R is mapping every point U in the real fields $Rn+$.

The existence of the utility function has been demonstrated (Zeng et al. 2006) (Chen et al. 1999) that it is monotone, continuous, differentiated. Several common binary explicit utility functions are as follows.

1   The alternative (linear) utility function $u(x_1, x_2) = a_1 x_1 + a_2 x_2$

2   The complementary utility function: $u(x_1, x_2) = \min\ (a_1 x_1 + a_2 x_2)$

3   The merging utility function: $u(x_1, x_2) = u(x_1) + u(x_2)$

4   The alternative (exponential) utility function: $u(x_1, x_2) = x_1^a + x_2^a$

5   Cobb - Douglas utility function $u(x_1, x_2) = x_1^a \bullet x_2^a$

When the total utility in the application uses multiplication rule, if any element is zero, the total utility value is zero. When it uses addition rule, if every element value in the total utility is independent, then each other can be linear compensated.

Cobb - Douglas utility function:

$$U(x) = x_1^{a_1} \cdots x_n^{a_n}$$
$$= \prod_{i=1}^{n} x_i^{a_i} \ \left(a_i \geq 0, \quad \sum a_i = 1 \right)$$

$$(1)$$

Here (1) the $U(x)$ is the utility, and the $x = (x_1, x_2, \cdots, x_n)$ is a vector of goods consumed. The utility consumer got is power product of all kinds of goods amount.

The utility theory we known from economics was established according to the principle of diminishing marginal utility and total utility maximization. Consumers buy goods in order to obtain the utility and they are willing to paying more for big marginal utility of goods. It means that consumers take a standard of the marginal utility as to pay for goods. The utility, in the same way, can be obtained from the different combinations of several articles. Some modern economists think that it is not necessary for consumers to feel the satisfactory degree of precise measurement, only know consumer's utility order through their preference for different commodities. All of these utility functions are presented many economic fields in the finance, the stock market or the insurance and are only not applied to an article or a book.

4. BELLMAN EQUATION

The general function of the utility to resolve a problem can be divided into optimization and control, expectation utility, preferences and risk, substitution and so on. Its function makes up elements of the value of the goods such as amount and increment as variable with the array, and parameter as variable coefficient with normalization. There are commodity, time and price in variable as unconstraint or constraint condition. The unconstraint condition is only to establish a utility function relationship and the constraint condition is to solve the function value. In the constraint condition the cost total is much application as variable condition. There are also many variables such as the utility total (U), scalar quantity (X), the number of similar products or the amount of costumer (x_n), price (p_n) and the constraint condition such as the total funds (w) and total time (T). They are established according to different application to the specific event in the utility function.

Bellman equation is often used of the multi-stage decision-making under uncertainty modeling. It was original as the engineering control theory in applied mathematics, and then became an important tool in economics theory that mainly solved the problem of optimal control theory within any time. It usually refers to the dynamic programming equation and associates with the discrete-time optimization (2) (See Wiki Bellman equation 2011a).

$$V(x_0) = \max_{\{a_t\}_{t=0}^{\infty}} \sum_{t=0}^{\infty} \beta^t F(x_t, a_t)$$

$$s.t. \quad a_t \in \Gamma(x_t), \quad x_{t+1} = T(x_t, a_t), \quad \forall t = 0, 1, 2, \cdots \tag{2}$$

Notice that $V(x_0)$ notation has been defined to represent the optimal value that can be obtained by maximizing this objective function subject to the assumed constraints. This function is the value function. It is a function of the initial state variable x_0, since the best value obtained depends on the initial situation.

Bellman equation is proposed the optimality principle to separate the current decision-making from the future decision-making. The stage and the state in the first term must constitute the optimal decision in the future, which reflects the basic idea of the dynamic programming method.

The continuous time optimization should require to partial differential equation, usually with the Hamilton-Jacobi-Bellman (HJB) equation (3) (See Wiki Constant elasticity of substitution 2011b).

$$V(x(0), 0) = \min_{u} \left\{ \int_0^T C[x(t), u(t)] dt + D[x(T)] \right\} \tag{3}$$

The $C[--]$ is the scalar cost rate function and $D[--]$ is a function that gives the economic value or utility at the final state, $x(t)$ is the system state vector, $x(0)$ is assumed given, and $u(t)$ for $0 \le t \le T$ is the control vector that we are trying to find.

The Constant Elasticity of Substitution (CES) is a property of some production function and utility function. It is a type of production function that displays constant elasticity of substitution. In other words, the production technology has a constant percentage change in factor (e.g. labor and capital) proportions due to a percentage change in marginal rate of technical substitution. The two factors (Capital, Labor) CES production function introduced by Solow and later made popular by Arrow, Chenery, Minhas, and Solow is equation (4):

$$Q = F \cdot \left(a \cdot K^r + (1 - a) \cdot L^r\right)^{\frac{1}{r}}$$

$$(4)$$

The Q is the output, the F is the factor productivity, the a is the share parameter, the K and the L are the primary production factors (capital and labor), the r is the elasticity of substitution (See Wiki Hamilton–Jacobi–Bellman equation 2011c).

When two variables become the proportion of the relative charges, there is an elasticity relation between variables. The CES has the nature of the production and utility function and means a specific type of aggregation function to combine two or more types of consumption, or a function relation among two or more species productive inputs into overall amount. And the CES utility function means the relation between the consumption amount and the utility for consumers of the elasticity of substitutive inconvenience.

Bellmen equation was applied to the resources allocation (Zhao 2004), assuming the A is the total number of some kind of economic resources, and assign it to the N ($N \ge 2$) economic units U_i ($i = 1, 2, \cdots, N$), assigned i to the amount of resources of economic units U_i is the A_i ($A_i \ge 0$), then the revenue L_i of the economic units U_i is the $L_i(A_i)$. If the maximum total Revenues $\tilde{L}^\sim N$

(A) are earned by economic resources A, the resources allocation will earn the maximum total revenues as follow equation (5):

$$\tilde{L}_N(A) = \max_{\sum_{i=1}^{N} A_i = A} [L_1(A_1) + L_2(A_2) + \cdots + L_N(A_N)]$$

(5)

The most optimal allocation of total resources is as follow equation (6):

$$\tilde{L}_N(A) = \max_{0 \le A_N \le A} \left[L_N(A_N) + \max_{\sum_{i=1}^{N} A_i = A} \{L_{N-1}(A_{N-1}) + \cdots + L_1(A_1)\} \right]$$

(6)

The project evaluation in the venture capital was established mathematical model (6) with Bellmen equation (Qian and Huang 2003), where the F is a value of investment option, the x_t is a current state variable for enterprise, the π is a profit function for an enterprise, the r is risk-free profits, the V is value to convert cash flows in the future for the project, the α is a expected growth rate of the cash flows for the project, the σ is a fluctuating rate of the cash flows for the project, the t is current time, the T is a expiration date of the investment option, and the I is investment costs. When an enterprise selects decision-making u_t at the length of time each stage Δt, the discount factor is $1/(1+r\Delta t)$, the enterprise has the maximal expected value with the current income and the future investment option (7).

$$\text{[illegible equation]}$$

(7)

When an asymmetric monopolistic competition model (8) was established with utility function, a set of different brands in the same type of products $(x_1, x_2, \cdots, x_n)$, the substitution parameter ρ between 0 and 1, it reflected the strength or weakness for diversity preferences of consumers (Weng and Chen 2006).

$$U = u\left[x_0, \left(\sum_i X_i^\rho\right)^{\frac{1}{\rho}}\right]$$

(8)

The early utility made a point of the economics such as consumption and the utility function was applied to the many economical fields. The current utility in our research is to use articles such as reading books in library collection. These utility functions are not adapted to the evaluation as a method and exploring new utility function is the need in order to evaluate books collected.

5. METHODOLOGY

The basic concepts of the utility management based on various expositions above all can be extracted from the utility (Say 1982) and the utility value (Zheng 2003) (Yu and Hu 2007) and the utility function (Zhang and Chen 2000) (Koksalan and Ozpeynirc 2009) and the utility theory (Farquhar 1984) (Dong 2007) as a management theory. It can be summarized mainly as follow: (1) the utility of resources or the product as a unit of measure; (2) the amount of using resources or the using product as evaluation; (3) determining the user's preferences through the measure values; (4) the new resource allocation or the article diversity to hedge against the risk through the measure values; (5) continuous time utility assessment. Its goal is to mine the potential usefulness of the article and to predict the degree of the article continual used and to reach the article full used and maximal effects.

Because book evaluation is one of important work in library management, utility function may be applied to the evaluation as a major method and become one part of utility management. Bellman equation in utility functions has the property of the multi-stage decision-making under uncertainty modeling and very adapts evaluating to collecting books and using books as a method in library management.

While the utility of the article in the utility management is a use role, the total utility can be separated into a basic utility and factors utility. The basic utility is an inherent use value of the article and it may get a theoretical use role through the market assessment. The factors utility is a real use role of the article and it is uncertainty caused by various factors. If an article was not properly managed, it might cause the total utility to reduce. Otherwise if it was maintenance well, its total utility would extend or increase. The total utility may predict a concrete article- how many to reduce or extend on the basis of actual usage data. Because each library has its own structure of library preservation, different characters of each librarian make books existed utility as soon as collected after consumed and had not presented consumed utility for the library. The books in library preservation have the basic utility such as knowledge, appreciation, conservation exhibit etc. and other utility as factors utility will present use data in loan, read, citation estimation and other. The utility management shows all practical use roles in total utility and can have a kind of book to play a different utility part in every region. On these grounds we become Bellmen equation into follow (9) as a new mathematical modeling of utility function for book evaluation:

$$U(x) = \max_{0<t\leq\infty} \left[U(x_0, y_0, t) \right]$$
$$+ \max_{1\leq t_e \leq T} \left\{ \sum_{j=1}^{n} \beta U \left[x_j^r(t_e), y_j(t), t \right] \right\}$$

(9)

Here the $U(x)$ is the value of optimal utility in formula (9); it is divided into the basic utility $U(x_0, y_0)$ at the first term on right side of formula and the factors utility $U(xrj, yj)$

at the second term on right side of formula. The basic utility of books had collected utility soon as preserved by library is reflected in books of the inherent value and heritage value of literature. The collected time $t=t_e$ is determined the state (x_0, y_0) of kinds of books which is a continuous function of time, that is, the longer is time, the greater is the basic utility. The factors utility of books is uncertain, and xrj is variables of variety library preservation, where the r is the constant of substitution elasticity, the $\beta<1$ is parameters converted or discount factor with use effect by readers. When time t_e and total preservation x are certain value, x_j is preservation variable at every preservation place, y_j is use variable of readers, they are a function of discrete random variables. The most optimal factors utility is to require the utility function in utility management. The factors utility may divide into collection-related utility such as borrowing, download within the network, guided reading, recommendation, communication, display and so on, and utility of nothing to do with the collection such as references, reviews, awards and so on. If $x = (x_1, x_2, ---, x_n)$ is vectors of the preservation, $|A|$ is a count matrix of kinds of the preservation and if $y=(y_1, y_2, ---, y_n)$ is vectors of the usage, $|B|$ is a count matrix of kinds of the usage.

In utility management amount of book utility is set $x = (x_1, x_2, \text{---}, x_n)'$ when $t = t_e$, then $y = (y_1, y_2, \text{---}, y_n)$ is all use vector when collected time $t = t_e$, utility time $T \geq t_e$ is use time of readers. The factors utility of books $U = U(xrj, yj, t)$

when the r is constant, $y_j = y_j(x, T),\ U = U\left[y_j(T), t\right]$. The factors utility predicted in the theory should have a fixed value U_0, and the utility management will present how to abridge or prolong theory utility U_0. Utility different $\Delta U = |U_0(x, y_0, T) - U(x_j, y_j, t)| \leq \varepsilon$ according to management effect. When $t = t_e$ and $x = \sum(ax_j),\ t_e < t \leq T$ and actual count of usage is $y_i(x)$, and y_0 is theory count of usage. When $y_0 \neq y_j$ then $0 < |y_j - y_0| < \delta$ is existence. Assuming that $S(y_i)$ is a use function, theory use function $S(y_0) = A,\ |S(y_0) - A| < \varepsilon$, then $\lim y_j \to y0 S(yj) = A$. Same $\lim\limits_{y_j \to y_0} U(y_j) = U(y_0)$ exist in the extreme value of the factors utility.

We can define some variables and derive some explicit functions from in (9). The methods of book evaluation are as follows:

- At the right second term of the formula (9) if the value of total factors utility is less than or equal to 1, the discount factor β is effective, and if it takes an arbitrary plurality, the discount factor β is ineffective. The discount factor as a utility weight can be artificially set or can also be obtained through use vector.

- At the right second term of the formula (9) the multivariate of the total use is simplified to mono variable that makes up function with time t, the amount of usage is not direct proportion to the amount of the preservation. Assuming it to be an exponential relationship:

$$y(x, t) = \left(Ae^{-\lambda x} + a\right)t \quad (A > 0,\ \lambda > 0) \tag{11}$$

- When the formula (11) is done the partial derivatives with time t, and we get the extreme value $Ae^{-\lambda x} + a = 0, x = -\ln(-a/A)/\lambda$ that attain the maximum use of the preservation. Here the amount of the preservation is $x = (x_1, x_2, ---, x_n)$, the n is places of the preservation, the λ is species, the x is a copy, and λx is the total copies of every species.

- After the right second term of the formula (9) is simplified, assuming the use variable is an amount of lend books or holding days of books, the utility function $U=U(y, t)$ is done the partial derivatives with time t, and we get an amount of loan books or holding days of books at every piece of time. If we need to solve the utility of different kinds of books, it is the ordinal utility that is the first, the second, the third, etc. to reflect the utility ordinal or grade. It is also to rank order method according to reader preference, that is, usually to use ranking list method.

- After the right second term of the formula (9) is done the partial derivatives with $U(x_j)$, we obtain $\dfrac{\partial U(x)}{\partial U(x_j)} = \dfrac{1}{\beta'}$, in which discount factor is an actual utility factor, it is the book usage factor (BUF) (Yan 2009b), and can be obtained by empirical data.

$$BUF = \frac{\partial U(x_j)}{\partial U(x)}$$
$$= \frac{\text{checkout number of one kind books}}{\text{total checkout number of same class books}} \tag{12}$$

- At the right second term of the formula (9) the multivariate of the total use is simplified to mono variable that makes up function with time t, book usage half-life (Yan and Wu 2011), is illustrated as (13):

$$r = \frac{dx}{dt} = kx^n, \qquad dt = -\frac{dx}{kx^n} \tag{13}$$

- When the $n=1$, $t = 0$, the total amounts of checkout books is x_0, $c=\ln x_0$, see (14):

$$\ln x - \ln x_0 = -kt, \quad \ln\frac{x}{x_0} = -kt, \quad x = x_0 e^{-kt} \quad T_{\frac{1}{2}} = \frac{\ln 2}{k} \tag{14}$$

- A principle of the utility maximization leads to exist in a possibility of the greatest use risk. The individual behavior to judge, that is a decision-making, is in order to obtain the maximum value of the expected utility, so called the preference, and not necessarily to gain the maximum benefits. If some use amount y and satisfaction degree u form utility function $u=u(y)$, it has the first derivative equal to zero and the second derivative be less than zero. When the maximum total utility is unchanged, some utility can exist in the use risk because every use variable consisted of every kind of utility which some small utility was ignored.

- The substitution role exists in the cardinal utility, with the r denoted a coefficient of elasticity of substitution, and the r is obtained

$$r = \frac{\Delta y}{y} \Big/ \frac{\Delta x}{x}$$

- the y and x denote different kind of preservation and Δy and Δx denote different kind of use preservation.

6. RESULTS

The data collection was from information management system of some university libraries and statistical results were also from the methods of the changing Bellman equation. We can get some aspects as follows.

6.1 Diversity judgment

It is an inevitable result of the natural development that the article diversity is the same as the biological diversity. The biological diversity is environmental needs for maintaining ecological balance and sustainable development, while the article diversity is the guarantee of human and environment for harmonious living together in a long period. The category diversity in library preservation can meet the needs of readers' choice, and judgment of the diversity could obtain through book usage factor calculated. We calculated some book usage factors about the management as keyword in library of China University of Mining and Technology (CUMT). See Table 1.

The utility management can express article diversity and their diversity using as total utility in utility function. Once data collected were calculated to ignore the smallest classes of utility, books exist in diversity short. The value of BUF in Table 1 is greater than or equal to 0.33 such as CLC (Chinese Library Classification) number G627, F2-42, R95, C93-7, F407.162.1, F407.162, X9-05 and F205 to judge single class or diversity short. We think the value of BUF should be smaller than or equal to 0.3 because of the article diversity at least 3 kinds.

6.2 Early warning for risk

We get the total amount from certain number of users using about different species in same class of articles in a certain period of time and call as all the utility U. The relationship is between a certain use time T and total utility U as the article time $T = t_n - t_0 + 1$, to establish numbers of usage X and time t with a function $X = X(t)$, then the area covered is total utility. However, the different purpose of articles may have different numbers of usage, which occupy different weighty. Use weighty a_i of different purpose is first obtained, and then we can get the total utility $U(x) = U(x_{11}, x_{22}, ---, x_{nm})$.

Table 1 Different BUFs of management's books in library of CUMT

Title	CLC number	BUF	Checkout times	Checkout times of same class
University management	G627	1.000000	5	5
"Management of National Economy" Self-counseling	F2-42	1.000000	1	1
Pharmacy Management	R95	1.000000	3	3
Western Management Famous Summary	C93-7	0.800000	48	60
Modern Management of the Coal Enterprises	F407.216.1	0.562500	9	16
Coal Industry Production Management	F407.162	0.454545	5	11
Security Management Principles	X9-05	0.389831	23	59
Resources Management	F205	0.351351	13	37
Modern Book Industry Enterprise Management	G231	0.272727	3	11
Education Management	G46	0.262626	29	110
Archives Management	G271	0.258741	37	143

$$\sum k_i = K, \quad \sum a_i = 1$$

$$(15)$$

Here K is amount of copies every kind. One of use weights becomes $a_{1j} >> a_{(n-1)j}$, while $U(x) \approx a_{11}x_{11} + a_{12}x_{12} + \cdots + a_{1m}x_{1m}$. Use duration is likely to the sum of all use time after articles into the management. This usage is determined by number of articles or use time. We know that use of book includes reading, citation, review, collection etc., data of which collected also have each different and the reading book (loan) is only main use or the greatest than all other use.

We have get the top ten kinds of books about economics and management on the ranking list of the sales from "*China Publishers*" (2010) and these books were collected at some university library and show Table 2.

In Table 2 the top books were not collected by all university libraries on the ranking list of the sales. There were 70 copies of every kind at some book store in Nanjing. Because data of citation from CNKI are all from the kind of book, data of reading are likely one library or some libraries at the region area.

There were 51 kinds of books collected by NUPT about CLC number TN711, network, by the end of 2009 and total valid lending times were 77. The part retrieval results show fellow Figure 1 by our program designed and the top five kinds of books are follow as Table 3.

Table 3 shows the first five kinds of books and 69 times of lending, occupying 90% of total times. Other books were used less and even neglected to use. The books, they mostly use, are a few of kinds and are published in recent 3–5 years. It is uncertain at book supple chain from publish, sell, collection and read.

Table 2 Books collected about economics and management at some university libraries

No.	Title	Publish date	Library name	Citation times
1	The World is Flat	2008.7	(1)(2)(3)(4)(5)(6)(d)	1734
2	Lang Xian-ping Said: the New Imperialism in China	2010.1	(1)(2)(3)(4)(5)(6)(d)	8
3	Currency War	2007.6	(1)(2)(3)(4)(5)(6)(d)	203
4	Currency War 2	2009.7	(1)(2)(3)(5)(6)(d)	10
5	Lang Xian-ping Said: the New Imperialism in China 2	2010.5	(1)(2)(3)(5)(6)(d)	
6	Detail Determining Success or Failure	2004.2	(1)(2)(4)(5)(6)(d)	690
7	Future Enterprise Road	2010.4	(1)(2)(4)(5)(6)(d)	1
8	Lang Xian-ping Said: Who is to Save China's Economy	2009.9	(2)(3)(5)(d)	
9	The Power of the Company	2010.8	(1)(2)(3)(6)(d)	5
10	Lang Xian-ping Said: Why is it so Hard Our Days	2010.9	(1)(3)(6)	

Note: *the (1) denotes library of People's University of China; the (2) denotes library of Central University of Finance and Economics; the (3) denotes library of Nanjing University; the (4) denotes library of Southeast University; the (5) denotes library of Nanjing University of Posts and Telecommunications (NUPT); the (6) denotes library of Nanjing University of Finance and Economics; the (d) denotes Duxiu Scholar of ChaoXing company in China.
*Statistical date: February 25, 2011.
*Citation of statistical date: March 9, 2011.

Figure 1 Part retrieval results of books about CLC number TN711.

Table 3 Contrast BUFs among different kinds in same class of books

Title	BUF	Checkout times	Total checkout times
Electrical Network Theory (Volume one)	0.311688	24	77
Electrical Network Theory (Volume two)	0.298701	23	77
Net Theory - Network Flow	0.181818	14	77
Diagrams, Network and Algorithms	0.064935	5	77
Microwave Network	0.038961	3	77

The library preservation has existed in the utility and the risk together. The presence of risk is often a potential element that undermines utility balance. There are two extreme aspects of very high and very low or never use amount. The high usage will be asked to continue to increase collections and to result in deformity collection phenomenon of additional a few kinds and many copies of one kind. The books, they used less or never used, keep in purchasing as a simple collection utility with no value of BUF. Books as collection may refer to high value of BUF in some university libraries. Our common goal will effectively configure and share with literature resources in regional area that commonly use books would collect in every university library and books of feature categories of profession or specialty would collect completely in different university library formed special collection.

6.3 Preference continuing

In many practical applications, the utility function can get an explicit function of multiple high-order equation group and extreme solution. For example in book usage half-life (BUHL), according to book classification, books belong to different classes, there are different kinds in the same class, and the same kind may has number of checkout times or checkout times, taking-out days, appointment times, citation times, book reviews, number of prints, number of editions and so on. If taking a category of books and checkout amount, according to defining BUHL, we take annual average x_0 of checkout books within three years that were same class into library. When the number of checkouts reduces to a half of the annual average, this year would be a half-life of using books, meaning a degree of aging books.

There were, in the same time, 284 kinds and 1101 copies with CLC number R94, pharmacy, in library of China Pharmaceutical University in 2004. Book usage half-lives of the classification up to April 2010 are as follows Table 4 and Figure 2.

6.4 Substitution role

Such as print books and electronic books in preservation resources exist in amount of different use and collection in different period, and the amount of use and collection print books was far greater than its e-books in the last century. But now the rate of print books between the amount of use and collection is quite some even lower than one of the e-books. In the future the electronic resources will be flooding in published market but re-use of print books will improve.

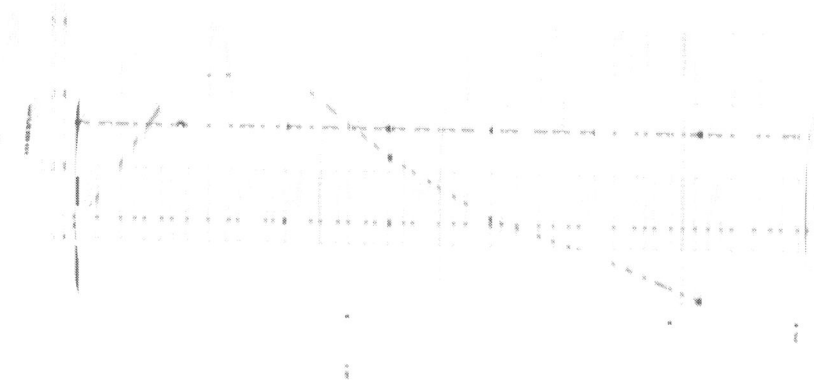

Figure 2 Book usage half-life of CLC number R94.

Table 4 Book usage half-life of CLC number R94

Year	2004	2005	2006	2007	2008	2009	2010
Checkout	71	196	201	132	82	54	18

When BUHL shows a peak again in 2011, the number TP of CLC exists in continual preference of readers. See Figure 3.

In library of NUPT, there were 800 thousand copies numbers of total library preservation and 506071 times of total checkout in 2005. There were 1 million copies numbers of total library preservation and 642622 times of total checkout in 2007. There were 19 literature database in Chinese and English and 224148 download pieces in 2005 and 37 literature database in Chinese and English and 149813 download pieces in 2007. Assuming that the letter y denotes print books and the letter x denotes electronic literatures. The constant of coefficient of elasticity of substitution r is follow:

$$r_{2005} = \frac{\Delta y}{y} / \frac{\Delta x}{x}$$
$$= \frac{\frac{506071}{8000000}}{} / \frac{224148}{19} \approx 0.6326/11797 \approx 5.362 \times 10^{-5}$$

$$r_{2007} = \frac{\Delta y}{y} / \frac{\Delta x}{x}$$
$$= \frac{\frac{642622}{10000000}}{} / \frac{149813}{37} \approx 0.6426/4049 \approx 1.587 \times 10^{-4}$$

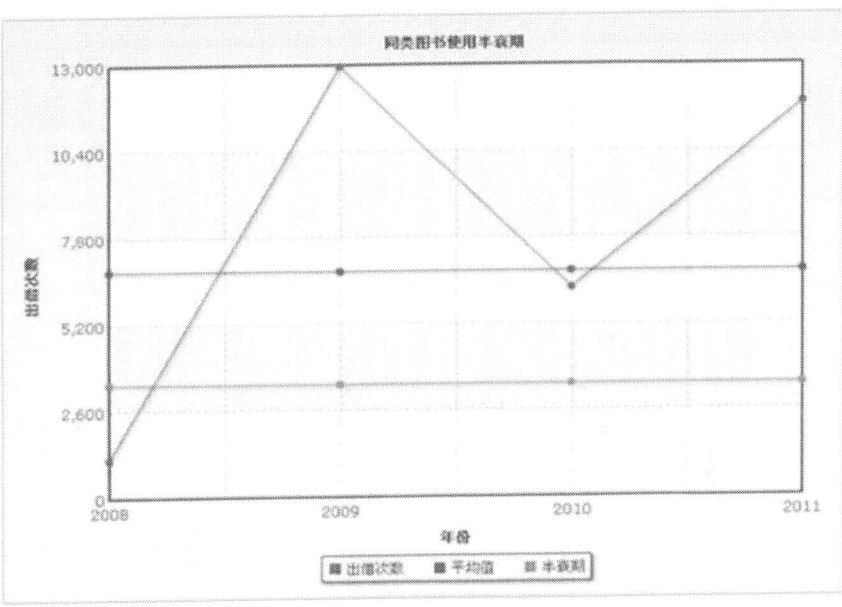

Figure 3 Book usage half-life of CLC number TP.

When $r_{2005} < r_{2007}$, it shown that print books in 2005 was less checkout numbers than in 2007, but reading print book in 2005 was more weight than downloading electronic literatures as compared with in 2007. The electronic literatures from 2005 to 2007 had not substituted the print books.

DISCUSSION AND CONCLUSIONS

The utility is no longer a preference for obtaining commodities but the preference for articles using. It means that a user was enthusiastic about the use of some kind of article. If the long-time or a lot of certain articles could be used, the user who loses interest would attempt to try new commodity and substitute for using article, and the amount of use would decline and total utility would have the maximum value. According to the definition of marginal utility to satisfy the final consumer commodity utility of that unit, so called total utility increment, there is the marginal utility for use of the article. It says that whether it is commodities or articles, its commodity property has been lost, only property of use remains. The utility role can be expressed by utility value and utility function. Only commodities are the economic utility from the view of consumer and articles are the management utility from the view of user.

With the amount of book publishing and its marked price have increased year by year, the library is difficult to maintain its previous method of procurement with the book procuring fund increased slightly. Also due to the collection in the form of electronic publication, nearly half of the literature procuring fund spends the literature databases to bring about same collection among university libraries. The utility in library management is a role of using book so we propose the utility management as a complement of the collection management in library. We hope to know whether readers read collected books or not through analysis results and make some suggestions.

Many early methods of book evaluation lack science and system unlike the journal impact factor. Many utility functions were defined as implicit function and applied to economics as the power function form of the logarithm and the exponential function of explicit function. There is not a utility function for book evaluation in all of these. The study has found Bellman equation can be required on the evaluation overall quality of books as measure model in library management. In our exploration we introduce the possibility of book evaluation utility function on the basis of library collections. We propose the book usage factor and the book usage half-life from changing Bellman equation as an evaluation method and to improve a method of traditional evaluation and assessment of books. A set of assessment methods based on utility management is able practically to evaluate the collecting books and forecast readers to obtain the knowledge and skills preferences.

If the utility is a degree of satisfaction for consumer to buy goods, the utility will be a degree of satisfaction for user to use amount of articles by managers. The mathematical modeling for the utility management needs to establish a functional relationship between articles managed and amount of articles used and is able to give the maximal utility through analyzing management factors. Users may give different results to articles used and all results will give social utility and economic utility. Then utility management is to evaluate the synthetic utility that issues from articles managed after these used. The kind of evaluation takes a method about management factors for utility and it will predict the possibility to give maximal utility with a variety of factors. From evaluating books we have got that:

1) Being clear the use value of the articles in utility;

2) Determining the changing Bellman equation as a utility function in utility management;

3) The book usage factor and the book usage half-life are an explicit function in the utility function;

4) Two explicit functions play a better role in the book evaluation of collection;

5) Utility management as collection management supplement.

The utility management applied to books as an attempt and it can bridge the gap between the library and book supply and also provide for a measure model to improve a method of traditional evaluation and assessment of books. It will provide an optimal path for the preservation of cultural resources and made full use of the current resources. The study on utility evaluation for books based on the theory of utility value contains designing the standard formula, which converted lending books to the factor of statistical data, and book usage factor for evaluation method for book utility, which judged utility value of the kind of book for reader to obtain knowledge, and book usage half-life, which predicted readers reading trend for some class of books, and the elasticity coefficient of substitution, which established early warning of using fluctuations between print books and e-books. All mathematical formulas above need a great amount of data to illustrate reliability and give quantitative evaluation. The book evaluation needs to be further research by a utility function like the coefficient of elasticity of substitution in the future.

Total utility sums to all various utility for one kind of book or one class of books, while derivative utility does not direct become part utility and is only to judge its maximum utility. The statistical results display that "ranking list" could not affect library collection but the risk was a lot of collected books being in idle. Many kinds of books lack of the diversity and are not satisfaction for reader need. A kind of book is not read in a university library but has many readers in other university library. This needs to build special library collection among regional university libraries and make books in idle be maximum utility. It is feasible of the evaluation books with the changing Bellman equation and its results proved that the varieties diversity of books collection could meet the needs of readers and achieve book species continuity. While there was risk of collecting books that were not used for a long time, the kinds of books could be used in other university library. These university libraries may consider regional resources sharing. We proposed a changing

Bellman equation as utility function and some results of the evaluation books were from its derivative function. However some results were lack of more data analysis like the constant of coefficient of elasticity of substitution because of more electronic literatures download in current now, so as above all are just as a reference method.

ACKNOWLEDGMENTS

This paper is one of research reports of the project that title is "Research on Book Evaluation Measure Model Based on Utility" and the project, No. 10YJA870028, was supported by the Research Foundation of Humanities and Social Sciences of Ministry of Education of the People's Republic of China.

COMPETING INTEREST

The authors declare that they have no competing interest.

AUTHORS' CONTRIBUTIONS

YB carried out the utility study, function results and written in the whole article. SF participated in the part of material collection and literature review about utility. YRQ participated in the part of data calculated and figures made with a computer. All authors read and approved the final manuscript.

REFERENCES

1. Bellman equation. 2011a. http://en.wikipedia.org/wiki/Bellman_equation

2. Constant elasticity of substitution. 2011b. http://en.wikipedia.org/wiki/Constant_Elasticity_of_Substitution#CES_production_function

3. Hamilton–Jacobi–Bellman equation. 2011c. http://en.wikipedia.org/wiki/Hamilton-Jacobi-Bellman_equation

4. Bhuvaneswari M, Seethalakshmi P: Application based utility adaptation with prioritized weight assignment strategy for minimizing delay in multimedia applications. International Journal on Computer Science and Engineering 2011, 3(4):1672-1680.

5. Chen YH, Sun CY, Jing R: Mathematical analysis of the utility function. Journal of Chongqing Teachers College (Natural Science Edition) 1999, 16(2):19-22. 52

6. Dong CS: Failure mode and effects analysis based on fuzzy utility cost estimation. International Journal of Quality & Reliability Management 2007, 24(9):958-971. 10.1108/02656710710826216View Article

7. Farquhar PH: Utility assessment methods. Manag Sci 1984, 30(11):1283-1300. 10.1287/mnsc.30.11.1283View Article

8. Feng SF: Utility function apply to finance. Journal of Beijing Polytechnic University. 2010, 19(1):119-121.

9. Fereidunian A, Zamani M, Lesani AH, Lucas C, Lehtonen M: An expert system realization of adaptive autonomy in Electric Utility Management Automation. J Appl Sci 2009, 9(8):1524-1530. 10.3923/jas.2009.1524.1530View Article

10. Ferne BH: Collection evaluation in the research library. Collect Build 1993, 9(3):33-37.

11. Jana S, Pavol K: Aggregation functions and personal utility functions in general insurance. Acta Polytechnica Hungarica 2010, 7(4):7-23.

12. Koksalan M, Ozpeynirci SB: An interactive sorting method for additive utility functions. Comput Oper Res 2009, 36: 2565-2572. 10.1016/j.cor.2008.11.006View Article

13. Li MG, Gao CL, Guo YL: Theory and method of utility management for project cost-effectiveness. Finance and Taxation 1998, 7: 30-31.

14. Lin YR: The study on utility function theory and apply to library. Mathematics in Practice and Theory 2005, 35(10):104-108.

15. Marx KH: Capital. Vol. 1 edition. Edited by: Guo D, Wang YA. Beijing, China: People's Publishing House; 1963:6.

16. Paul AS, William DN: Economics. the Fourteenth Version. Beijing, China: Beijing Economic College Press; 1996:150.

17. Qian ST, Huang ZY: Study on the risk investment projects to assess the quantitative model based on the ROV method. Quantitative & Technical Economics 2003, 10: 49-52.

18. Say JB: A treatise on political economy. Edited by: Chen FS, Chen ZH. Bejing, China: The Commercial Press; 1982:59.

19. Tan P, Guo L, Wand J: Annuity market hindrance based on Epstein-Zin utility function. Sci-Technology and Management 2006, 1(40–41):44.

20. Weng J, Chen LS: A monopolistic competition model based on a two level CES utility function. Journal of Central China Normal University (Nat Sci) 2006, 40(3):447-451.

21. Xiao XM: The analysis about the location and the management of utility of electronic reading room of university library. Investigation and Research 2003, (12):18-20.

22. Yan B: Statistical application of book availability. Libr J 2009a, 28(7):31-34.

23. Yan B: Statistics of body usage in library collection and its application. Library Management; 2009:86-89.

24. Yan B, Wu XQ: Application of utility management based on article usage. In 2011 International Conference on Management Science and Engineering (18th). Rome. Italy; 2011:174-180.

25. Yan B, Xu H: Outdated electronic resources of the utility management. Global Journal of Management and Business Research 2010, 10(6):36-39.

26. Yu TS, Hu SP: A review of the labor theory of value and utility value unified. Econ Rev 2007, 3: 3-9.

27. Zeng Y, Geng L, Zhang P: The mathematical analysis of utility function. Journal of Hubei University of Technology 2006, 21(1):47-49. 70

28. Zhang B: Modeling design of the books' circulating avail in university library. Journal of Modern Information 2009, 29(2):136-137. 141

29. Zhang YT, Chen HY: Utility function and optimization. Beijing, China: Science Press; 2000.

30. Zhang X, Wang L: Study on utility management base on the library union of the core value. Journal of Information 2005, 11: 131-133.

31. Zhao JM: The method of Bellman Equation about function problem. Study Progress on Mathematics, Mechanics, Physics and High-tech. Conference on Mathematics, Mechanics, Physics and High-tech in China 2004, 10(10th):64-66.

32. Zheng KZ: On the value of the objective utility theory: reconstruction of the micro-foundation of the political economy. Jinan, China: Shandong People Publishing House; 2003.

CHAPTER 3

Deposition of suspended fine particulate matter in a library

Jiří Smolík[1*], Ludmila Mašková[1,2], ,Naděžda Zíková[1,3],
Lucie Ondráčková[1], and Jakub Ondráček[1]

1 Institute of Chemical Process Fundamentals ASCR, Prague, Czech Republic.
2 Institute of Environmental Studies, Faculty of Science, Charles University in Prague, Prague, Czech Republic.
3 Department of Meteorology and Environmental Protection, Faculty of Mathematics and Physics, Charles University in Prague, Prague, Czech Republic.

ABSTRACT

To analyse deposition of fine particulate matter (PM) on book surfaces we put twelve bunches of cellulose filters on a free shelf of the National Library in Prague, exposed them for three, six, nine, and twelve months to indoor air and analysed them after each period by Scanning Electron Microscopy (SEM) and Ion Chromatography (IC). Results showed that fine particles were deposited predominantly on the surface of the top filter but partly also on the surfaces of inner filters. It indicates fine particles penetrated between filters. The penetration and deposition of particles was also modelled as Brownian diffusion between two parallel filters. The model prediction demonstrated that fine particles penetrate between filters, with the depth of penetration limited by parallel diffusional deposition on filter surfaces. This is in qualitative agreement with SEM and IC investigations. The results show that beside the top part fine PM can deposit onto all available surfaces of books.

KEYWORDS

Fine particulate matter Deposition Brownian diffusion National library in Prague

1. INTRODUCTION

Airborne particles deposited on cultural heritage artefacts have many negative effects. Beside soiling and abrasion of surfaces particles can also cause material deterioration by chemical reactions. Ultrafine atmospheric particles, penetrating indoors from the outdoor environment, contain soot and organic matter from traffic that are hygroscopic and effective for transport of acids. Fine particles consist of secondary organic matter and ammonium sulfate, ammonium nitrate, and sometimes sulfuric acid. Coarse particles, formed predominantly by resuspended dust contain crustal elements and in the indoor environment sometimes alkaline particles emitted from concrete structures [1, 2].

There are several studies reporting particle deposition measurements in museums and other buildings with works of art or historical artefacts [3–13]. The deposition has been determined using vertical and horizontal collection plates or filters, attached to the walls or horizontal surfaces with subsequent analysis of amount and chemical composition of deposit by various techniques. The results, though somewhat diverse, have several common features: a) the amount of deposit on the horizontal surfaces is larger than on vertical surfaces, b) the composition of deposit corresponds to the composition of particulate matter (PM) suspended in the indoor air, and c) particles deposited on vertical surfaces are formed frequently by soot, organic matter and ammonium sulphate. The rate of submicron particle and sulfate deposition onto a vertical surfaces measured inside five Southern California Museums were in a good agreement with theoretical prediction [14].

Dry deposition is considered to occur by a combination of Brownian and eddy diffusion and gravitational settling [15, 16] where prevailing deposition mechanism depends on the particle size. Coarse particles are deposited on upward-facing surfaces by gravitational settling and fine particles predominantly by diffusion on surfaces of any orientation. In principle the submicron particles can penetrate by diffusion also between books and even into the gaps between pages and thus can be deposited on the inner surfaces of books. To test this hypothesis we examined deposition of particles on paper filters located on the free shelf of the library.

2. EXPERIMENTAL

The deposition experiments were performed during one year comprehensive study of indoor air quality in the Baroque Library Hall of the National Library in Prague, that included measurements of gaseous pollutants, size resolved particle number concentrations, and size resolved chemical composition of indoor PM. The results showed that the average concentrations of size fraction 100 nm - 1 µm of the indoor PM was of the order 10^3 particles/cm³. The IC analyses revealed that the major water-soluble norganic component of this fraction was ammonium sulfate with maximum concentration centred at about 300 nm [17].

To investigate deposition of fine PM we placed twelve bunches of ten cellulose Whatman filters No. 542 (circles, 70 mm), fixed in open Petri dishes, on a free shelf of the library (Figures 1 and 2). Each bunch contained ten filters, which were gently loosed to increase the distance between them. Whatman filters were chosen since due to low blank were suitable a) for subsequent analyses of deposit and b) for study of possible effect of deposited PM on degradation of cellulose, as main component of paper. In the experiment bunches of filters represented books, laying on the shelf.

Figure 1 Exposed Whatman filters.

Bunches were exposed for three, six, nine, and twelve months and examined after each period by Scanning Electron Microscopy (SEM) and Ion Chromatography (IC).

3. PARTICLE PENETRATION AND DEPOSITION

To estimate penetration of particles between filters and subsequent deposition on inner surfaces we modelled transport of particles by Brownian diffusion between two parallel discs (Figure 3) put in environment with constant particle number concentration n_0.

Figure 2 Detail of exposed Whatman filter.

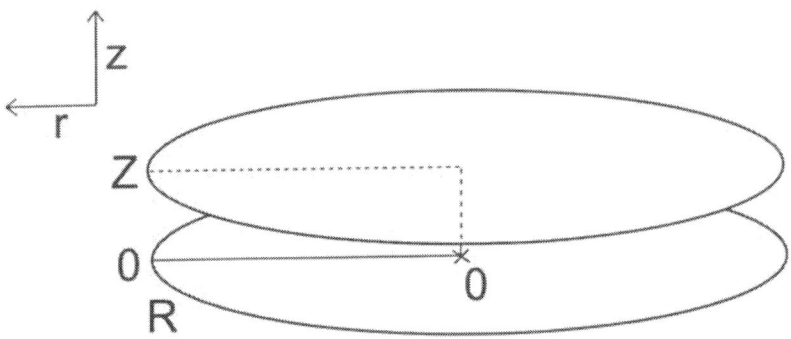

Figure 3 Diagram for two parallel discs.

The particle concentration was described by the equation

$$\frac{\partial n}{\partial t} = D\left(\frac{1}{r}\frac{\partial n}{\partial r} + \frac{\partial^2 n}{\partial^2 r} + \frac{\partial^2 n}{\partial^2 z}\right)$$

(1)

where n is the particle number concentration, t is the time, r and z are the radial and axial distances, respectively, and D is the particle diffusion coefficient [18]

$$D = \frac{kTC_c}{3\pi\eta d_p}$$

(2)

where k is the Boltzmann's constant, T is the temperature, C_c is the slip correction factor, η is the air viscosity, and d_p is the particle diameter. The slip correction factor C_c is given by

$$C_c = 1 + \frac{\lambda}{d_p}\left[2.34 + 1.05\exp\left(-0.39\frac{d_p}{\lambda}\right)\right]$$

(3)

where mean free path λ for air at standard conditions is 0.066 μm. The solution of the equation (1) with boundary conditions

$$
\begin{array}{llllll}
n & = & n_0 & for\ r = R, & z \geq 0 & and\ t \geq 0 \\
n & = & 0 & for\ r \geq 0, & z = 0 & and\ t \geq 0 \\
n & = & 0 & for\ r \geq 0, & z = Z & and\ t \geq 0 \\
\dfrac{\partial n}{\partial r} & = & 0 & for\ r = 0, & z \geq 0 & and\ t \geq 0
\end{array}
$$

(4)

gives the concentration gradient $\partial n/\partial z$ at filer surfaces $z = 0$, $z = Z$ and using Fick's law the rate of particle deposition per unit area of surface

$$J = -D\frac{\partial n}{\partial z}\Big|_{z=0, z=Z} \tag{5}$$

The equation (1) was solved numerically for particle sizes d_p = 10, 100, and 1000 nm, gap width $Z = 1$ and 5 mm, and constant particle number concentration $n_0 = 1.10^3$ particles/cm³. An example of steady state concentration profile of 1000 nm particles in 5 mm gap is shown in Figure 4.

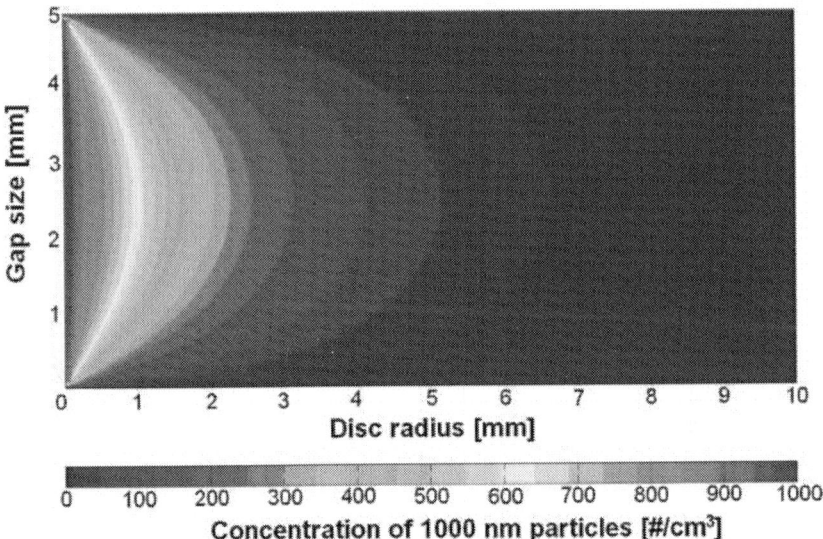

Figure 4 Concentration of 1000 nm particles between two parallel discs with gap 5 mm.

As can be seen the concentration of particles rapidly decreases with increasing distance from the edge due to fast deposition by axial diffusion. To estimate the deposition we calculated amount and mass of particles (assuming spherical particles of unit density) deposited on inner surface during one year of exposition. (i.e. $J.t$). The results showed that particle penetration and deposition depends on particle size and width of the gap, with the depth of the penetration limited by parallel diffusional deposition on filter surfaces. Smaller particles penetrated faster and deeper resulting in higher particle number concentration of deposited particles but higher mass was transported by larger particles.

4. SCANNING ELECTRON MICROSCOPY

Surfaces of the top (first) and internal (second) filter were analysed by Scanning Electron Microscopy (Quanta 450, FEI, USA). The samples were coated by gold and viewed with the microscope at 20 kV. Typical example of particles deposited on the top filter is shown in Figure 5. At high magnification particles of different appearance and size were distinguished: spherical and irregular particles, larger aggregates of smaller particles, and fibres with sizes between about 50 nm to 50 μm. Some particles were also analyzed by using SEM-EDAX. Large irregular particles contained usually crustal elements Si, Al, and Ca indicating mineral origin. Several smaller spherical particles contained C and O and occasionally showed some instability under the electron beam indicating biological matter. Examination of internal filters showed that the particles were deposited on small area extending about 1 mm from the filter edge, that qualitatively corresponds to the model prediction. However, direct comparison was difficult, since model differs from the real arrangement: filters fixed inside shelf on Petri dishes were not exactly parallel and gaps between filters were not constant (Figure 2), surface and edge of filters had a fibrous character (Figure 5), that differs from the smooth surface assumed in modelling, and indoor aerosol particles were polydisperse with highly time variable concentrations.

5. ION CHROMATOGRAPHY

Results from the parallel study [17] showed that ammonium sulfate formed up to 60% of mass of water-soluble part of indoor submicron PM. Since ammonium sulfate is stable compound, it is considered as a tracer of ambient fine aerosols that penetrate indoors [19, 20]. In this study sulfate was used as a marker for deposited particles.

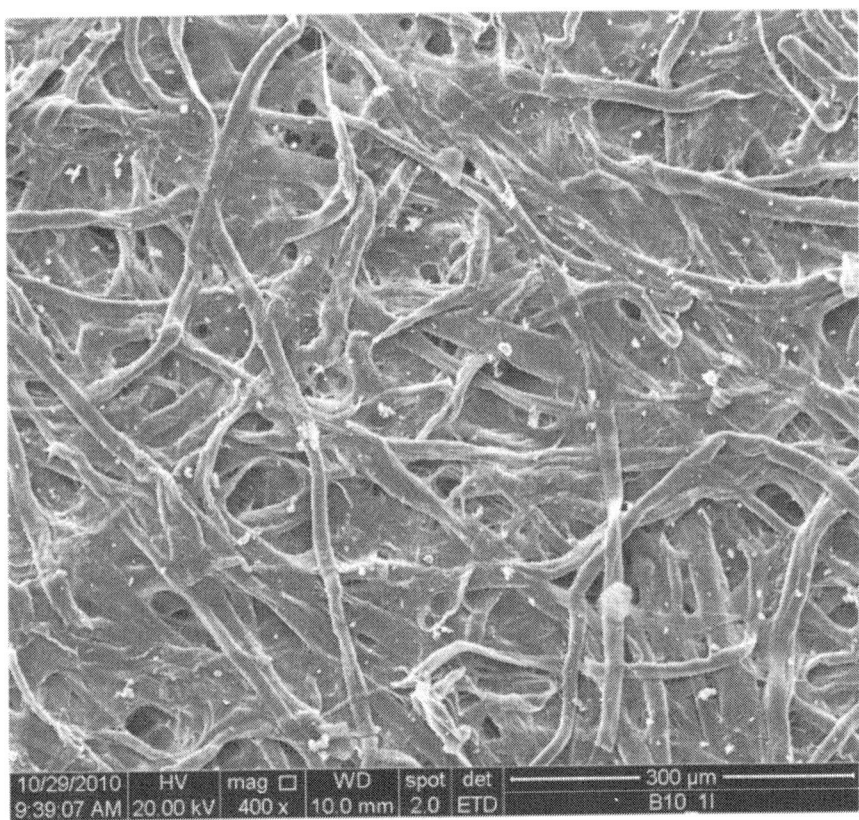

Figure 5 SEM photograph of exposed Whatman filter.

Exposed filters were analysed after each period by Ion Chromatography (IC). Since the bunches were not exactly planar but slightly convex (see Figure 2) the downward-facing surface of the last filter could be exposed as well. Thus, we analysed the top (first), internal (second) and bottom (tenth) filter from two bunches after each period. The analyses were provided using the setup by Watrex Ltd., Czech Rep., with columns Watrex IC Anion II 10 μm 150 × 3 mm for anions and Alltech universal cation 7 μm 100 × 4.6 mm for cations. The conductivity detector used in this setup was a SHODEX CD-5. The samples were extracted by 10 ml of ultrapure water with conductivity 0.08 μSm^{-1} (Ultrapur, Watrex Ltd.) for 0.5 h using ultrasonic bath and 1 h using a shaker. The filtered extracts were then analysed for anions and the next day for cations. The sample solutions for cation analysis were stored in the fridge prior the analysis.

Results of analyses, corrected for the blank value, are shown in Figure 6. As can be seen, sulfate concentrations on all filters increased with time. Substantially higher sulfate concentrations were found on the top filter predominately due to larger exposed area. These results were confirmed by statistical analysis (Wilcoxon test, $p < 0.05$), which showed significant differences only between the top and other two filters. A non-parametric test was chosen, because the Shapiro-Wilk test rejects the hypothesis of normal distribution for the bottom filter at $p < 0.05$. The results clearly show, that submicron particles penetrated between filters with transport governed at least to the downward-facing surface (Figure 2) of the bottom filter by diffusion.

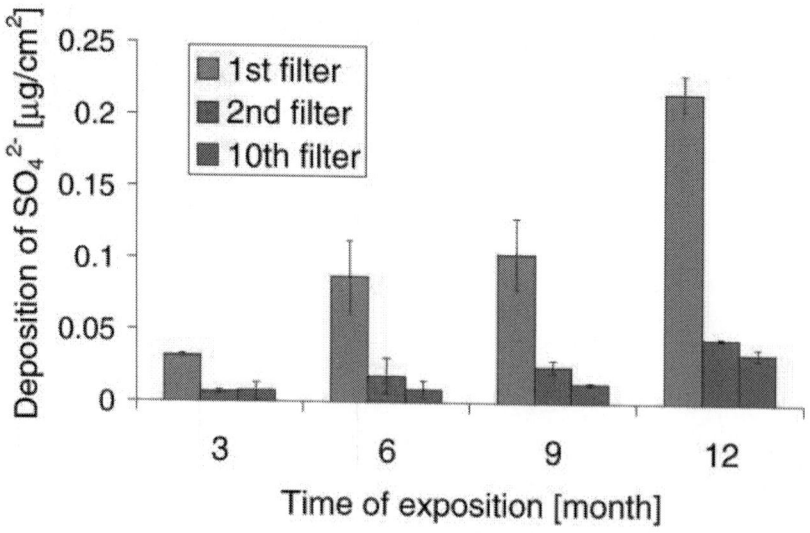

Figure 6 Concentration of sulfate ion deposited on the top (first), internal (second) and bottom (tenth) filters during three, six, nine, and twelve months long exposition. Average values obtained from two parallel bunches, the bars represent the standard deviations.

6. CONCLUSIONS

To test if the indoor submicron particles can penetrate by diffusion into gaps between books or even between pages of books, we exposed twelve bunches of Whatman filters to the indoor air fixed in open Petri dishes on a free shelf of the library. Bunches were exposed for three, six, nine, and twelve months and analysed after each period by Scanning Electron Microscopy (SEM) and Ion Chromatography (IC). The penetration and deposition of particles has been modelled assuming Brownian diffusion between two parallel discs, The simple model showed that the particle penetration and deposition depends on particle size and width of the gap, with the depth of penetration limited by parallel diffusional deposition on filter surfaces. Smaller particles penetrate faster and deeper resulting in higher particle number concentration of deposited particles, but higher mass is transported by larger particles. The results of modelling have been qualitatively confirmed by Ion Chromatography using sulfate as marker for deposited particles. The results show that beside the top part fine PM can deposit onto all available surfaces of books.

ACKNOWLEDGEMENT

This work is a part of research project No. DF11P01OVV020 supported by the Ministry of Culture of the Czech Republic.

COMPETING INTERESTS

The authors declare that they have no competing interests.

AUTHORS' CONTRIBUTIONS

JS prepared the manuscript. LM performed analyses. NZ performed mathematical modeling. LO performed PM measurements. JO performed PM measurements. All authors read and approved the final manuscript.

REFERENCES

1. Nazaroff WW, Ligocki MP, Salmon LG, Cass GR, Fall T, Jones MC Liu HIH, Ma T: **Airborne particles in museums.** J Paul Getty Trust 1993. Publication of the Getty Conservation Institute, available online: *http://www.getty.edu/conservation/publications_resources/ pdf_publications/pdf/airborne.pdf*

2. Hatchfield PB: Pollutants in the museum environment. London: Archetype Publications Ltd; 2005.

3. Ligocki MP, Liu HIH, Cass GR: **Measurements of particle deposition rates inside southern California museums.** Aerosol Sci Technol 1990, **13:**85–101.

4. Nazaroff WW, Salmon LG, Cass GR: **Concentration and fate of airborne particles in museums.** Environ Sci Technol 1990, **24:**66–77.

5. Christoforou CC, Salmon LG, Cass GR: **Deposition of atmospheric particles within the Buddhist cave temples at Yungang, China.** Atmos Environ 1994,**28**(12):2081–2091.

6. De Bock LA, Van Grieken RE, Camuffo D, Grime GW: **Microanalysis of museum aerosol to elucidate the soiling of paintings: case of the Correr museum, Venice, Italy.** Environ Sci Technol 1996, **30:**3341–3350.

7. Brimblecombe P, Blades N, Camuffo D, Stuarto G, Valentino A, Gysels K, Van Grieken R, Busse H-J, Kim O, Ulrich U, Wieser M: **The indoor environmet of a modern museum building, the Sainsbury centre for visual arts, Norwich, UK.** Indoor Air 1999, **9:**146–164.

8. Camuffo D, Brimblecombe P, Van Grieken R, Busse H-J, Sturaro G, Valentino A, Bernardi A, Blades N, Shooter D, De Bock L, Gysels K, Wieser M, Kim O: *Indoor air quality at the Correr museum, Venice, Italy.* Sci Tot Environ 1999, *236:*135–152.

9. Gysels K, Deutsch F, Van Grieken R: *Characterisation of particulate matter in the Royal museum of fine arts, Antwerp, Belgium.* Atmos Environ 2002, *36:*4103–4113.

10. Gysels K, Delalieux F, Deutsch F, Van Grieken R, Camuffo D, Bernardi A, Sturaru G, Busse H-J, Wieser M: *Indoor environment and conservation in the Royal museum of fine arts, Atwerp, Belgium.* J Cult Herit 2004, *5:*221–230.

11. Worobiec A, Samek L, Spolnik Z, Kontozova V, Stefaniak E, Van Grieken R: *Study of the winter and summer changes of the air composition in the church of Szalowa, Poland, related to conservation.* Microchim Acta 2007, *156:*253–261.

12. Worobiec A, Samek L, Krata A, Van Meel K, Krupinska B, Stefaniak EA, Karaskiewicz P, Van Grieken R: *Transport and deposition of airborne pollutants in exhibition areas located in historical buildings-study in Wavel castle museum in Cracow, Poland.* J Cult Herit 2010, *11:*354–359.

13. Nava S, Becherini F, Bernardi F, Bonazza A, Chiari M, García-Orellana I, Lucarelli F, Ludwig N, Migliori A, Sabbioni C, Udisti R, Valli G, Vecchi R: *An integrated approach to asses air pollution threats to cultural heritage in a semi-confined environment: The case study of Michelozzo's Courtyard in Florence (Italy).* Sci Total Environ 2009,*408*(6):1403–1413.

14. Nazaroff WW, Ligocki MP, Ma T, Cass GR: *Particle deposition in museums: comparison of modelling and measurement results.* Aerosol Sci Technol 1990, *12:*332–348.

15. Lai ACK, Nazaroff WW: *Modelling indoor particle deposition from turbulent flow onto smooth surfaces.* J Aerosol Sci 2000,*31*(4)*:*463–476.

16. Nazaroff WW, Cass GR: *Mass-transport aspects of pollutant removal at indoor surfaces.* Environ Int 1989, *16:*567–584.

17. Andělová L, Smolík J, Ondráčková L, Ondráček J, López-Aparicio S, Grøntoft T, Stankiewicz J: *Characterization of airborne particles in the Baroque Hall of the National Library in Prague.* e-Preservation Science 2010, *7:*141–146.

18. Hinds WC: Aerosol technology, properties, behavior, and measurements of airborne particles. 2nd edition. New York: John Wiley & Sons, Inc; 1999. Chpt. 7 Brownian Motion and Diffusion

19. Slanina J, ten Brink HM, Otjes RP, Even A, Jongejan P, Khlystov A, Waijers-Ijpelaan A, Hu M, Lu Y: *The continuous analysis of nitrate and ammonium in aerosols by the steam jet aerosol collector (SJAC): extension and validation of the methodology.* Atmos Environ 2001, *35:*2319–2330.

20. Pszenny A, Fischer C, Mendez A, Zetwo M: *Direct comparison of cellulose and quartz fiber filters for sampling submicrometer aerosols in the marine boundary layer.* Atmos Environ 1993, *27A:*281–284.

CHAPTER 4

The Problem of Subject Access to Visual Materials

Heather P. Jespersen[1], John Kresten Jespersen[2]

[1]*Image Classification Specialist, John Carter Brown Library at Brown University*
[2]*Technical Services Supervisor Curry College*

ABSTRACT

This article discusses the problem of giving subject access to works of art. We survey both concept-based and content-based access by computers and by indexers/catalogers respectively, as well as issues of interoperability, database and indexer consistency, and cataloging standards. The authors, both of whom are trained art historians, question attempts to mystify fine art subject matter by the creation of clever library science sys- tems that are executed by the naive. Only when trained art historians and knowl- edgeable catalogers are finally responsible for providing subject access to works of art, will true interoperability and consistency happen.

KEYWORDS

Subject access; Cataloging; Visual information; Vocabularies; Subject control

1. THE SPECIAL CASE OF ART

The problem of the bibliographic control of non-print media found its twin focus in content-based access by computers and in concept-based access by human catalogers and indexers over thirty years ago. In 1971 Oscar Firschen and Martin Alvin Fischler wrote their seminal *Describing and abstracting pictorial structures* in a major text on pattern recognition and optical data processing. This work clearly foreshadows the content- based access by computers of visual data that has been a pressing concern during the 1990s. In 1972 Bette C. Acuff and Joan Sieber-Suppes at Stanford University wrote *A Manual for Coding Descriptions, Interpretations, and Evaluations of Visual Art Forms*, an early work in the concept-based access of visual materials by human catalogers and indexers. The scope of this paper deals with the problem of giving subject access to art images. It is primarily concerned with the describing, cataloging, and indexing of works of art. It is not concerned with popular or news imagery, nor with scientific imaging or imagery. But this paper is concerned with how to give access to both iconographical subject matter and to more formal visual processes such as shape, texture, and color. Art history is not driven by subject matter alone, although most of the studies of concept-based access have focused on the iconological theories of Erwin Panofsky. Formalism, or the concern with the purely visual qualities of a work of art, is best treated in our opinion by computer content-based access as qualified by the human cataloger s intervention.

A recent Australian study of the problem has found that a mix of concept and content, of the iconographic and the for- mal, yielded the best search results for a web based search engine (Lu, Williams, & You, 2000). Since a work of art is a combination of both subject matter and more formal painterly characteristics, we take an approach to subject access that considers both dynamic elements. Moreover, since the advent of post-modern scholarship, issues of gender, society, multicultural- ism, and colonialism transcend simple subject matter and make for a vastly more complex discipline in art history and therefore for greater intellectual demands on the cataloger and indexer.

Translating visual works into a verbal form is obviously difficult (Shatford, 1984, 1986, 1994; Markey, 1984, 1986, 1988; Svenonius, 1994). But this translation is just as obviously doable, or art history would not be an academic discipline with roots in the Renaissance artist biography of Vasari. Samuel Taylor Coleridge links the visual to language in his essay On Poesy or Art. There he writes that First, there is mere gesticulation; then rosaries or wampum; then picture-language; then hieroglyphics; and finally alphabetic letters. These all consist of a translation of man into nature, of a substitution of the visible for the audible (Coleridge, 1965, p. 266). Picture language could be ornament, or any kind of representation of reality that is in effect beginning to substitute the visible for the audible by communicat- ing information as ideas, concepts, and words to the beholder. These visual ideas grow increasingly abstract as the structure of language progresses from gesture (as for example in dance or mime) to the alphabet as the written word. Aleph, for example, the first letter of the alphabet in Phoenician, Greek and Roman, originates as the stylized head of an ox. Letters are them- selves originally pictures; hieroglyphs are more pictures than words, but nevertheless convey information of a verbal and audible kind. In thinking about the problem of providing subject access to visual information, it is best to keep in mind the unalterable connections between the verbal and the visual and not to create dichotomies between the two where none effectively exist.

One of the major problems that arise when cataloging visual items concerns the issue of interoperability. More important even than uniformity of interfaces is the adoption of common vocabularies that will populate those interfaces. Users can adapt to different methods of searching, but they will probably learn certain search terms-the ones that bring results-and it is not fair to the users to have an infinite variety of terms or descriptors. However, since a variety of users needs must be met (Spencer, 1995), it is generally acknowledged that limiting the vocabulary is difficult and may even be counter productive. At the moment, a number of nationally-recog- nized controlled vocabularies for concept-based subject access and utilities for authority control and cataloging standards exist.

2. A REVIEW OF SOLUTIONS

Perhaps the first systematic attempt to create subject control and access for works of art was begun by Prof. Charles Rufus Morey, an iconographi- cally-oriented art historian at Princeton University, in 1917. He began the Index of Christian Art as a card file, one part of which was a listing of sub- jects. Initially only available at Princeton, three more copies were located in the U.S. and in Europe. The material covered is works of art in all media up to A.D. 1400 (with some recent expansion to 1600) that appears in a Christian context. Computerization of the Index began in 1991, and it is now partially available online at <http://ica.princeton.edu/>. The Internet version lags behind the vast card file and contains 20,000 records and 60,000 images as compared to the card file s 200,000 reproductions. The online version has 150 searchable categories. Three-quarters of the subject terms have been mapped to ICONCLASS notations. The database is available by subscription only, but most major academic libraries subscribe.

Most similar to Princeton s Index in its approach and content is ICON-CLASS, a subject indexing tool developed by Professor Henri Van de Waal beginning in the 1950s, left unfinished at his death in 1972, and completed by others in the 1980s. ICONCLASS has been available on line by subscrip- tion since 1995 (http://www.iconclass.nl). As its name implies, it deals only with iconography. The categories are created *a priori*, that is abstractly rather than in response to a need to describe a particular work of art. The 24,000 definitions of objects, persons, events, situations, and abstract ideas are hierarchically arranged. Each category is assigned an alphanumeric nota- tion that can function in a way similar to a Dewey Decimal number. The cre- ators of ICONCLASS assert that this form of notation makes linguistic bar- riers to the use of the system irrelevant. Karen Spencer counters that the alphanumeric notations are frustratingly meaningless and have to be extract- ed and constructed from a seventeen-volume set of books. The numbers are interlocking, making it difficult to add new categories. The hierarchical sys- tem does, however, make it relatively easy to collocate similar subjects.

The Library of Congress Thesaurus for Graphic Materials was developed by Elisabeth Betz Parker for LC s Prints and Photographs Division. TGMI is a list of 6300 subject terms (http://www.oloc.gov/rr/print/tgm1). TGMII contains terms for genre and physical characteristics. Strings can be created as in LCSH using nationality, geographic, chronological, and topical subdi- visions. Terms are provided for activities, objects, types of people, events and places. Proper names for objects, people, events, and places are not included so that, for example, one can find saints but not Saint Jerome. Neither does it provide terms for art historical or iconographical concepts. Abstract concepts such as might be exemplified in allegorical prints and editorial cartoons are represented. Indexers are guided by scope notes, and the arrangement of terms is hierarchical. It has been pointed out by Jorgensen (Jorgensen, 1999) that because of the scope of LC s collection, LCTGM contains a preponderance of terms relating to American history and institutions.

The Art and Architecture Thesaurus, originally developed by the J. Paul Getty s Art History Information Program (AHIP) in 1979 to provide access to architecture slides, is a hierarchical and faceted list of terms that can also be used to create a kind of string. Terms are fully defined and notes on usage are included. The AAT does not include terms for iconographical themes. Its seven facets are arranged from abstract to concrete concepts, and there are thirty-three sub-facets. Since it was developed to describe works of art, it tends to neglect people, events, and activities in favor of the physical attrib- utes of objects. Jorgensen mentions that a study has found the AAT of vary- ing usefulness to the naive user. Nevertheless, the AAT seems to be gaining in popularity, perhaps because of its breadth of coverage in comparison to the LCTGM, which is, after all, oriented toward graphic materials.

The utility for authority control of artists names, the Union List of Artists Names (ULAN) is also created by the J. Paul Getty Trust. Institutions submit forms of names as they find necessary. Each name is presented on line to the cataloguer in all its known variants, with one chosen as the preferred form. This is a conveniently accessible, but rather bare bones approach. Almost no information is given about the artist that would allow the cataloger to decide whether, in fact, artist X is the one with whom he/she has to do. Better in this last respect, but existing only in paper and CD-ROM (since 1995) formats is the traditional resource, the Allgemeine

3. KUNSTLER LEXIKON BY THIEME-BECKER.

The other cataloging utility developed by the Getty is the Thesaurus of Geographic Names (TGN) (http://www.getty.edu/ research/tools/ vocabulary/ tgn/about.html). It consists of one million names of places that have been contributed by participants. The names of continents, nations, cities, and physical features in their modern and historical forms are included. Each place is given a unique number and a considerable amount of information, both geographical and historical, is given about it. Its position in the hierar- chy is shown, relationships to other places are given, as are geographical coordinates, notes, sources for the data, and place types (state capital, etc.). Its name in English, other languages, in history, and in natural and inverted order is given. The preferred form is the one commonly used by the local population of the place, but a preferred English form is also given. It is a very satisfying and even interesting resource to use.

Standards or schemata for image cataloging have been developed by the Visual Resources Association (VRA) and the J. Paul Getty Trust. The VRA s system allows both the object being cataloged, usually a slide, digital file, or some other reproductive medium, to be described along with the original work that it represents. This is a very complete and informative sys- tem that works especially well for slide libraries. The VRA itself asserts its commitment to advocating standards for documentation of and access to images. The VRA Core Categories, now in version 3 (http://vraweb.org/vra- core3.htm), are considered a starting point, a set of basic fields, that encour- age interoperability but also allow flexibility and customization to fit the description of a particular collection. The VRA recommends the use of the Getty s controlled vocabularies and authority utilities.

The Getty s Categories for the Description of Works of Art is just that, a system for describing original works rather than reproductions, although sur- rogates can also be described. Its 231-plus fields allow for an exhaustive description of a work including its history, condition, and relationships to other works (http://www.getty. edu/research/ institute/ standards/cdwa). It was developed to serve professionals and scholars. Core categories are indicated. On its website, the Getty provides mapping or crosswalks to other systems, including MARC. MARC itself, of course, has been expanded to accommo- date new kinds of media, but has done so primarily by means of the addition of subfields. It is still sometimes a difficult judgment call to determine where the varieties of information attached to an image fit.

CIMI, originally know as the Consortium for the Computer Exchange of Museum Information, was founded in 1990 to encourage the use of stan- dards for description and access among members of the museum communi ty. It is concerned with the uniformity of electronic standards. The CIMI Standards Framework was published in 1993. The Framework advocated SGML as a means of structuring information and ANSI Z39.50 as the stan- dard for search and retrieval. A test project, Cultural Heritage Information Online (CHIO), was designed to test these standards. Member databases were marked up using SGML, and a common set of access points and a tag- ging guide were created. This initial project focused on the technological aspects of information sharing rather than on the commonality of fields and their contents.

Recently, the Library of Congress has initiated a new schema for the Digital Library Federation. The Metadata Encoding and Transmission Standards (METS) is a standard for encoding descriptive, administrative, and structural objects within a digital library, using XML (http://www. loc.gov/standards/mets).

Two major organizations that have more or less successfully attempted to create on-going image databases large enough to test and establish universal standards are AMICO and the Library of Congress American Memory and Prints and Photographs Online Catalogue (PPOC) projects. The Art Museum Image Consortium (AMICO) is made up of a group of institutions with art collections who all contribute to a common database as a way of making the educational use of museum material possible. Access is by subscription and is open to all educational institutions. Searches can be done by either con- cept- or content-based methods, making many types of searches successful. The site has recently been converted to Luna Imaging s Insight software which is very powerful for searching and for manipulating images once they are found. In fact, Insight is well on its way to becoming the preferred image database software here and abroad. Insight supports VRA, CDWA, MARC, and locally devised standards.

In 1998, the Library of Congress launched the site containing its enor- mous collection of prints and photographs, thus making it available to researchers online. At the moment, American Memory contains seven mil- lion images, and the PPOC contains 13.6 million (with one half online). As stated above, the LCTGM was created to provide access to the PPOC. Since the LCTGM is based on literary warrant, it is limited in usefulness for cata- logers dealing with non-American materials or materials outside certain his- torical confines.

The amount of writing that has been done in the last fifteen years con- cerning the problems involved in cataloging images is overwhelming. Yet, the concept- and content-based schemas and standards have still to be universally adopted. There seems to exist the conviction that uniformity is necessary, simultaneously with the conviction that freedom is desirable. Hence the existence of the situation among Luna Imagi s clients, each of whom is free to choose its own standards and controlled vocabularies. Luna hosts the sites of numerous major institutions and is encouraging simultaneous access to multiple collections and cross collection searching while support- ing each of their standards and providing software allowing each institution to create its own authority files.

It is true that libraries have a century of cooperative cataloging behind them, and that each museum, slide library, or other cultural institution has gone its own way because, supposedly, of the unique objects they possess. It is true that the physical characteristics of an artwork require more detailed description than do those of a book. But is it true that the description of the subject of a work of art is more complex than that of a book? Does its subject matter necessarily defy verbal description? Only in the cases of non- representational works is this somewhat true, and in these cases the name of the artistic movement to which the work belongs could stand in for a subject, since in Modernism and even Post-Modernism the two are contigu- ous. Also, even though each institution may hold unique works, there is overlap among the artists represented. Many museums own different but similar works by Frank Stella or Eva Hesse, and cataloging information relating to subject access could be shared just as it is for books.

4. CONTENT-BASED ACCESS AND CONCEPT-BASED ACCESS

Searching databases for purely formal and visual properties is already being done on content-based retrieval systems. These types of searches, for qualities of line, shape, texture, and color are pertinent to artists and to sci- entists alike, and someday the two systems, concept-based and content- based, should merge. At the moment there are a number of systems online or about to emerge. The most notable is IBM s Query by Image Content search engine. The QBIC system enables users to scan a visual database for color percentages, color layout, and textures. The State Hermitage Museum in St. Petersburg, Russia is the most famous patron of this particular content-based retrieval system. It claims that QBIC allows the user to search for artwork using tools that the artist would use (http://www.hermitagemuseum.org/ fcgi-bin/db2www/ qbicSearch.mac/qbic?selLang=English). By creating a virtual grid of colors, searches in the system match the required grid to other images in the Hermitage database.

Another successful online content-based retrieval system is VisualSEEk, created by John R. Smith under the supervision of Prof. Shih-Fu Chang at the Image and ATV Lab at Columbia University. The demo system at the Columbia University website (http://www.ctr. columbia. edu/ VisualSEEk) has an archive of 12,000 images that also includes videos. Searches are con- ducted by specifying both color and spatial locations on a grid. New search- able features recently added to the system include texture, shape, motion, and embedded text. Yet another system, or approach, is Toshikazu Kat s Query by Visual Example. While this system dates to before 1990 when it was first described by Kato, it seems most useful today as software in retrieving Trademark logos for graphic designers and probably lawyers. QBVE functions on the basis of retrieving all images in a database that look similar to the one in question. Finally, there is Visual Information Retrieval (VIR). These systems are on the verge of providing significant user help in tracking down appropriate visual information, especially in the complex area of videos (Gupta and Jain, 1997). The real beneficiaries of these systems are likely to be the scientists and health care professionals, but VIR might also benefit the art historians who are searching for appropriate research images without knowing in advance what those images should be, especially in the realm of Modern and Post-Modern art whose subject matter is often difficult to discern.

For art that has readily identifiable subject matter, the AAT and LCTG-MI can provide a controlled vocabulary equal to the task of providing sub- ject access to images just as LCSH does for books. Since both are based on literary warrant (unlike ICONCLASS), they are flexible. Both also address the physical characteristics of objects. Since that is the case, it is hard to understand why institutions assert that vocabularies peculiar to themselves are necessary. Within MARC one can indicate which vocabulary is being used, and one can even simultaneously use more than one. Use of these vocabularies within the MARC format would aid interoperability greatly.

Studies have been done of how much consistency can be achieved among catalogers of images, and a surprising degree can be achieved at the basic level of naming things in an image (Panofsky s first iconographic level) (Markey 1984, 1988; Shatford, 1986). The research that has been done on this has usually involved naive catalogers. In order to identify subject matter at more symbolic levels, a cataloger with subject training is necessary. Many studies employ naive catalogers, but why? Is it presumed that books are best cataloged by naive catalogers? When subject matter is difficult to determine, is the item given to the least experienced cataloger? Pictures are thought about and spoken about in words. They are created, because their creators want to express a concept visually. Why should con- cept-based subject access applied by a trained, experienced person not work as well for pictures as it does for books?

Critics of concept-based access to visual information claim that it is too time-consuming and therefore too expensive (Hsin-Liang Chen and Rasmussen). On reflection, it does appear that indexers for visual materials have too often not been art historians or subject specialists, and this fact alone will have upset the consistency of the results. We would argue that for consistent, economical results, trained subject specialists should be employed for the indexing of visual information. Art historians are trained in the methodology of art history and know the issues that are most relevant to potential art history searches. Terms such as metapicture, simulacrum, intertextuality, appropriation, commodification, historicism, essentialism, or remediation (Roberts, 2001) are the products of art historical scholarship and would be familiar to an indexer or cataloger trained in art history. However, a naive indexer would be baffled by such terms and what they rep- resent. Subject access provided by art history professionals would be the most efficient way to further unify the various indexes to visual collections of art around the world.

It has been nearly twenty years since Karen Markey Drabenstott wrote her first study on subject access to visual resources. Thanks to her study and to those of many others the problems of cataloging visual materials began to be discussed and understood. Between that time and the present the utilities and large scale projects discussed in this article have created and used the appropriate standards for image cataloging and have begun to provide a con- siderable degree of interoperability.

The authors would like to thank David P. Miller of Curry College for reading the article and providing criticism and encouragement.

REFERENCES

1. Markey, K.   Interindexer consistency tests: A literature review and report of a test of consistency in indexing visual materials, *Library and Information Science Research*, 6 2 (1984) 155-177.

2. Shatford, S. Describing a picture: A thousand words are seldom cost effective, *Cataloging & Classification Quarterly*, 4 4 (1984) 13-30.

3. Shatford, S. Analyzing the subject of a picture: A theoretical approach, *Cataloging & Classification Quarterly*, 6 3 (1986) 39-62.

4. Barnett, P. J. An art information system: From integration to interpretation, *Library Trends*, 37 2 (1988) 194-205.

5. Brilliant, R. How an art historian connects art objects and information, *Library Trends*, 37 2 (1988) 120-129.

6. Van der Wateren, J. F.   Achieving the link between art object and documentation: Experiences in the British Architectural Library, *Library Trends*, 37 2 (1988) 243-251.

7. Giral, A. At the confluence of three traditions: Architectural drawings at the Avery Library, *Library Trends*, 37 2 (1988) 232-242.

8. Markey, K Access to iconographical research collections, *Library Trends*, 37 2 (1988) 154-174. Besser, H.   Visual access to visual images: The UC Berkeley image database project, *Library Trends*, 38 4 (1990) 787-798.

9. Keefe, J. M. The image as document: Descriptive programs at Renssalaer, *Library Trends*, 38 4 (1990) 659-681.

10. Petersen, T. Developing a new thesaurus for art and architecture, *Library Trends*, 38 4 (1990) 644- 658.

11. Roddy, K. Subject access to visual resources: What the 90s might portend, *Library Hi Tech*, 9 1 (1991) 45-49.

12. Small, J. P. Retrieving images verbally: No more key words and other heresies, *Library Hi Tech*, 9 1 (1991) 51-60, 67.

13. Greenberg, J. Intellectual control of visual archives: A comparison between the Art and Architecture Thesaurus and the Library of Congress Thesaurus for Graphic Materials, *Cataloging & Classification Quarterly*, 16 1 (1993) 85-101.

14. Bell, L. A. (1994). Gaining access to visual information: Theory, analysis and practice of determining subjects. Retrieved February 26, 2003 from the World Wide Web:http://www.ilpi.com/artsource/ bibliographies/bellbib.txt.

15. Casey, D. D. Scouting new horizons: An annotated bibliography introducing subject access in visual image databases, *Illinois Libraries*, 76 (1994) 240-242.

16. Shatford Layne, S. Some issues in the indexing of images, *Journal of the American Society for Information Science*, 45 8 (1997) 583-588.

17. Svenonius, E. Access to nonbook materials: The limits of subject indexing for visual and aural lan- guages, *Journal of the American Society for Information Science*, 45 8 (1994) 600-606.

18. Spencer, K. (1995). Authority and vocabulary control in image collections. Student paper. Retrieved February 26, 2003, from the World Wide Web:http://sunsite.berkeley.edu/Imaging/ Databases/ Fall95papers/kspencer.html.

19. Armitage, L. H. & Enser, P. G. B. Analysis of user need in image archives, *Journal of Information Science*, 23 4 (1997) 287-299.

20. Gupta, A. & Jain, R. Visual information retrieval, *Communications of the ACM*, 40 5 (1997) 70-79.

21. Rasmussen, E. M. Indexing images, *Annual Review of Information Science and Technology (ARIST)*, 32 (1997) 169-196.

22. Sundt, C. The quest for access to images, *Advances in Librarianship*, 22 (1998) 87-106.

23. Arms, C. R. Getting the Picture: Observations from the Library of Congress on providing online access to pictorial images, *Library Trends*, 48 2 (1999) 379-409.

24. Forsyth, D. A. Computer vision tools for finding images and video sequences, *Library Trends*, 48 2 (1999) 326-355.

25. Heidorn, P. B. Image retrieval as linguistic and nonlinguistic visual model matching, *Library Trends*, 48 2 (1999) 303-325.

26. Hsin-Liang Chen & Rasmussen, E. M. Intellectual access to images, *Library Trends*, 48 2 (1999) 291-302.

27. Jorgensen, C. (1999). Image indexing: An analysis of selected classification systems in relation to image attributes named by naive users. Retrieved March 7, 2003, from the World Wide Web: http://www.oclc.org/research/publications/arr/1999/jorgensen/index. htm.

28. Sandore, B. Progress in visual information access and retrieval. *Library Trends*, 48 2 (1999). Retrieved January 23, 2003, from Omnifile database on the World Wide Web:http://vnweb. hwwilsonweb.com.

29. Djeraba, C., Bouet, M., Briand, H., & Khenchaf, A. Visual and textual content based indexing and retrieval, *International Journal of Digital Libraries*, 2 (2000) 269-287.

30. Giral, A. Digital image libraries: Technological advancement and social impact on the teaching of art and architectural history, *INSPEL*, 34 1 (2000) 9-21. Retrieved January 23, 2003 from Omnifile database on the World Wide Web: http://vnweb.hwwilsonweb.com.

31. Gordon, A. S. Browsing image collections with representations of common-sense activities, *Journal of the American Society for*

Information Science and Technology, 52 11 (2001) 925-929. Retrieved January 23, 2003, from the World Wide Web:http:// www3. interscience.wiley.com/cgi-bin/fulltext/85008785/.

32. Greenberg, J.   A quantitative categorical analysis of metadata elements in image-applicable metadata schemas, *Journal of the American Society for Information Science and Technology,* 52   11 (2001) 917-924. Retrieved January 23, 2003, from the World Wide Web: http://www3.interscience.wiley.com/cgi-bin/fulltext/85008948.

33. Jorgensen, C., Jaimes, A., Benitez, A. B., & Chang, S-F. A conceptual framework and empirical research for classifying visual descriptors, *Journal of the American Society for Information Science and Technology,* 52 11 (2001) 938-947. Retrieved January 23, 2003 from the World Wide Web:http://www3.interscience.wiley.com/cgi-bin/fulltext/85008946.

34. Lu, G., Williams, B., & You, C. An effective World Wide Web image search engine, *Journal of Information Science,* 27 1 (2001) 27-37.

35. Roberts, H. E. A picture is worth a thousand words: Art indexing in electronic databases, *Journal of the American Society for Information Science and Technology,* 52 11 (2001) 911-916. Retrieved January 23, 2003 from the World Wide Web:http:// www3. interscience. wiley.com/cgi-bin/fulltext/85008708.

36. Tam, A. M. & Leung, C. H. C.   Structured natural-language descriptions for semantic content retrieval of visual materials, *Journal of the American Society for Information Science and Technology,*52 11 (2001) 930-937. Retrieved January 23, 2003, from the World Wide Web: http://www3.interscience.wiley.com/cgi-bin/fulltext/85008715.

37. Wallace, M. C. The science and art of online research in the fine arts: A process approach, *Searcher,*9 8 (2001) 36-44. Retrieved January 23, 2003, from Omnifile database on the World Wide Web: http://vnweb.hwwilsonweb.com.

38. Thompson, M. C. Taking up the digital challenge: Image digitization projects at the Bibliotheque nationale de France, *Art Libraries Journal*, 27 3 (2002) 7-12.

39. White, L. Interpretation and representation: The who, why, what, and how of subject access in muse- ums, *Art Documentation*, 21 1 (2002) 21-2. Retrieved January 23, 2003, from Omnifile database on the World Wide Web: http://vnweb.hwwilsonweb.com.

40. Zinkham, H. Pitching pictures: the Prints & Photographs Online Catalogue at the Library of Congress, *Art Libraries Journal*, 27 3 (2002) 18-25.

CHAPTER 5

MExiCo: A Library for Managing Multimodal Data Collections

Peter Menke*, Philipp Cimiano

Project X1, SFB 673, Universität Bielefeld, Postfach 10 01 31, 33501 Bielefeld, Germany

ABSTRACT

We present MExiCo (Multimodal Experiment Corpora), a library specifically designed for the planning, modeling, management and analysis of multimodal corpora resulting from experiments, consisting of video and audio recordings as well as transcription and annotation documents.

1. INTRODUCTION

When multiple scientific disciplines join forces for large interdisciplinary projects, often opportunities to gain novel insights go along with clashes of theories, methods, and best practices. This is also the case for the Collaborative Research Centre (CRC) 673 "Alignment in

Communication", situated at Bielefeld University. It investigates the communicative phenomenon of alignment, as proposed by Pickering & Garrod (2004). Here, linguists, psychologists and computer scientists collaborate in 13 research projects, covering areas such as experimental psychology, theoretical linguistics, phonetics, cognitive science, and artificial intelligence. While 12 of these projects perform studies, experiments and simulations (involving both humans and artificial agents), the 13th project X1 provides support and infrastructure for the others. Its goal is to provide applications, libraries and data formats for the diversity of data collections created and maintained by the other projects. In this article, we present MExiCo, a library specifically designed to model the different flavours of data collections created by the different disciplines in such an interdisciplinary project. We give an overview of the major challenges, describe how MExiCo is structured and designed, and give some examples for everyday usage of the library.

Table 1. Research projects in the CRC 673 and the related disciplines.

Key	Title	Disciplines
A1	Modelling Partners	AI, psycholinguistics
A3	Dialogue and group dynamics	semantics, psycholinguistics
A4	Alignment of situation models	cognitive computer vision, psycholinguistics
A8	Constructions: Bridging the gap between syntax and semantics	theoretical linguistics
B1	Speech-gesture alignment	AI, linguistics
B4	Rhythm and timing in dialogue	phonetics, AI
B5	Anticipatory alignment in dialogue	psycholinguistics, neurolinguistics
B6	Understanding alignment from misalignment	clinical linguistics, cognitive robotics
C1	Interaction space	AI, cognitive robotics
C2	Communicating emotions	psycholinguistics, clinical linguistics, robotics
C4	Adaptive alignment in human-robot cooperation	human-machine interaction, cognitive robotics
C5	Alignment in in AR-based cooperation	human-machine interaction, cognitive robotics
X1	Multimodal alignment corpora	text technology, computational linguistics

2. THE CRC 673 AND ITS DATA COLLECTIONS

The research projects inside the CRC 673 (see Table 1 for a short summary) come from different disciplines, and base their work on different theories. As a consequence, also different methods are used. While many projects perform studies and experiments dealing with communication, those studies differ in what phenomena they investigate, and how they represent them. While phoneticians might be interested in convergence of rhythm in dialogue (Wagner, Inden, Malisz, & Wachsmuth, 2012), gesture researchers might look for completely different phenomena, such as an increasing similarity of gesture shapes in the course of a dialogue (Bergmann & Kopp, 2012). As a consequence, numbers and types of primary data (some research questions do not require video recordings, while others, especially gesture research, rely on multiple recordings from different perspectives to capture gesture details) as well as secondary data differ. Especially, there are numerous methods for creating speech transcripts, each focusing on a different subset of aspects (phonetic accuracy, orthographic assimilation, syntactic, assimilation, representation of prosodic features, representation of synchronous speech, etc.). Depending on the discipline and theory, some perform their analyses on the entirety of data (video and audio recordings along with transcripts and annotations) while others confine themselves to using only transcripts as soon as they are available.

As a consequence, numerous software applications and data formats gained currency in linguistics and its neighbouring disciplines, each filling a niche defined by theoretical background, by choice of methods and by design and nature of already existing data material:

1. Data formats defined for and used by transcription and annotation software, such as Praat (Boersma & Weenink, 2001), Transcriber (Barras, Geoffrois, Wu, & Liberman, 2001), Elan (Brugman & Russel, 2004; Wittenburg, Brugman, Russel, Klassmann, & Sloetjes, 2006), Anvil (Kipp, 2001), EXMARaLDA (Schmidt & Wörner, 2005; Schmidt, 2002) and its successor Folker (Schmidt & Schütte, 2010), and other, outdated formats;

2.   XML-based data formats for the representation of text and text-based (or single-timeline based) annotation, such as the TEI types (TEI Consortium, 2008), annotation graphs (Bird & Liberman, 2001), or the NITE object model (Evert et al., 2003);

3.   plain-text-based transcription formats, mostly from the context of conversation analysis, e.g., CHAT (MacWhinney, 2000), GAT (Selting, Auer, & Barth-Weingarten, 2009), or HIAT (Ehlich & Rehbein, 1976).

While researches have a background of theories, methods, and data and file formats they grew accustomed to, there is also need for exchange of such formats when it comes to interaction between researchers and projects. This could be the case when researchers intend to re-evaluate their hypothesis on the basis of the data collection of another researcher or project, or if multiple data collections are supposed to be unified, for instance, in order to form a resource large enough to serve as an adequate basis for lexicon creation, or for machine learning techniques. However, such an exchange becomes difficult when data formats are not compatible. Regularly, one-time converter scripts are written that make data sets compatible. Often, however, such a conversion is unidirectional in the sense that resulting data cannot be mapped back to the original data collection. If both data sets are used as bases for subsequent operations, branchings occur, with the consequence that data sets are even more difficult to re-unify at a later stage (cf. Lier et al., 2012).

Project X1 has the goal to provide libraries and tools dedicated to the reduction of such problems. One of them is the MExiCo library which is described in detail in the remainder of this article.

3. THE MEXICO (MULTIMODAL EXPERIMENT CORPORA) LIBRARY

The Multimodal Experiment Corpora (MExiCo) library has been designed to express the structure of a corpus or data collection based on structured sets of dialogues, conversations, and experiments involving dialogical behaviour. Its goal is to guide and support researchers during all stages in the lifecycle of such a data collection:

1. Structure and design of an experiment and its resources can be planned and organized with the design-related components, resulting in a design structure that serves as a kind of blueprint for the next phase, the actual experimentation phase.

2. During experimentation, MExiCo can provide checklist-like features to help experimenters keeping track of participants, variables, and resources. MExiCo can provide information whether all required elements of a collection are present and in a correct state or file format.

3. In the post-experiment phase, similar mechanisms are available to assist transcribers and annotators with data and file management. Missing tiers, layers, or tracks in annotation files can be identified, and pending tasks can be enumerated. Also, MExiCo provides basic mechanisms for the preparation and evaluation of agreement calculations of multiple annotations of the same phenomenon.

4. For analyses, data sets can be queried, transformed and exported into different file formats suitable for the major statistics software systems.

5. Finally, MExiCo provides corpus representations suitable for publication of whole corpora, sub-corpora, or single components therein, preferably using RDF and concepts of Linked Data (Chiarcos, Nordhoff, & Hellmann, 2012). This also includes the provision of metadata representations to specialised search engines and data harvesting systems, as, among others, the Virtual Language

Observatory (Van Uytvanck, Stehouwer, & Lampen, 2012). This procedure enables these systems to add entries for those resources to their search index, thus making them available in queries.

MExiCo models its data in two main levels of granularity (see Figure 1a): The macro level is for the representation of whole corpora and their major components, like entries for participants, variables and files or resources. For several types of resources (with transcription and annotation files leading the way), their inner structure is also modeled, thus providing access to tiers(which might be called" layers" or " tracks"

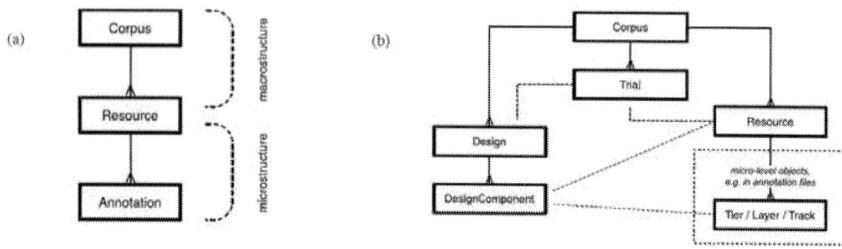

Figure 1. (a) Macro- and microstructure of MExiCo. (b) Simplified class diagram of MExiCo. Some classes (among them *Participant* and *Variable*) were omitted for the sake of clarity and readability.

3.1 Macrostructure: Corpora, their design and their components

The macro level of the library has a *Corpus* as its top-level element (see Figure 1b). A corpus consists of two groups of sub-elements: a design part and a resource part.

In the *design part*, the structures, numbers, and types of resources are defined. This part acts as a template or schema to which later resources and other actual data units can be assigned. A *Design* object models what properties one perfomance of an experiment is supposed to have: It defines the number and type of *Participants* and *Variables*, as well as the number and types of *Resources* that can be associated with one performance. Those specifications are given in *DesignComponent* objects.

As an example: The Bielefeld SaGA corpus (Lücking, Bergmann, Hahn, Kopp, & Rieser, 2010) is a homogeneous corpus where all performances follow a single design. This design calls for the following constellation of raw or primary data: three video recordings, each from a different perspective, one audio recording, and diverse body tracking data recordings. As secondary data, speech transcripts, annotations of part of speech and lemma information, and a detailed annotation of gestures is required. In MExiCo, this SaGA design would be modeled by creating a single *Design* object and then adding *DesignComponents* to it (for each video recording, the audio recording, and so on). Later, such a design can be used to check files for correctness and completeness (see below).

The performances, runs, or instances (or what else you prefer to call them) of an experiment are modeled as *Trial* objects. These can be associated with allocations of Participants and Variables in order to express who was involved, and which values the relevant variables have for a particular instance. The SaGA corpus has 25 trials, each involving two different participants (resulting in a total of 50 participants for the whole study).

Finally, a corpus consists of actual data. These are contained in *Resource* objects, typically present as files on a file system (other kinds of resources are possible in principle, such as non-digitised recordings on tapes, printed and hand-written transcripts, etc.). These resources are associated with other corpus components, such as trials and design components. This gives information about the particular dialogue the resource belongs to, and its type according to the underlying design (for instance, to express that a particular resource contains a video recording from the top-down perspective).

There are different *MediaTypes* a resource can belong to. Among them are *Video, Audio,* and *Annotation.* For annotation documents (which includes speech transcripts) there is special support in the micro-level part of MExiCo. Often, one annotation document contains multiple tiers, tracks, or levels of different kinds of data. Those should be associated separately to different design components. Also, computational access to single data elements of such files is necessary for queries, analyses and data transformations.

3.2 Microstructure: Annotation files and their internal structure

The top-level element of the microstructure part is an *AnnotationDocument* (which is a subtype of *Resource*). An annotation document contains a set of *Items,* which stand for single annotations, or equivalent objects. These items can be grouped together in *Layers* (which, depending on the system, might also be called tracks, tiers, or groups). Additional elements model scales from reality (for instance, timelines or spatial coordinates) and metadata information about the annotator, creation date, etc.

For each annotation file format used in the CRC 673 we developed import and export routines to and from our central MExiCo file format. One of the advantages of a single central file format is compliance to the DRY principle ("don't repeat yourself"): Each operation has to be designed and implemented only once, for the MExiCo data structures. Analyses of *all* supported third-party formats can then be performed by using the already existing import routine for the particular format.

File formats currently supported are: Praat ShortTextGrid and TextGrid files, Elan annotation documents, Anvil annotation documents, and (with limited support) EXMARaLDA documents.

Since the different third-party formats sometimes differ drastically, it is not always possible to perform exports to all file formats (for instance, when a third-party format does not support a certain type of hierarchical structuring of annotations present in a document). In order to get a grip on the plethora of supported and unsupported features of the individual third-party formats, a type and constraint system is being designed for the MExiCo library that is supposed to identify and handle such conversion problems. One of its goals is to provide solutions if an export or a conversion cannot be performed without loss. As an example, if a target format does not support complex annotation values, the user could be asked whether the data should be coded in an auxiliary format (lists and attribute value structures could be coded using JSON, the Java Script Object Notation). An alternative would be to spread information into multiple tiers: An annotation consisting of a list of word form, part of speech tag and lemma could be separated into simple-valued annotations on three different tiers. Hierarchical information could be expressed using a coding scheme in an additional tier, and so on.

4. APPLICATION

The MExiCo library is currently being separated from the application that first used it, the Phoibos corpus manager (Menke & Mehler, 2011), a web-based application for the management and department-internal sharing of corpus resources. It is implemented in the programming language Ruby, a second implementation in Python (which is undoubtedly more popular among linguists) is in progress.

MExiCo can be used in different ways: It can be used and included as a library in scripts and programs running locally, or it can be included in server-side applications (as in Phoibos), and accessed using standard web protocols such as HTTP(S). Also, the way how corpora and their components are stored can vary depending on the use case: Local folders and files are the default, but it is also possible to store frequently used data in a relational database. We plan to support remote file access (again, via standard web protocols) in addition.

MExiCo is already used actively in progamming courses, and in several Master projects in Bielefeld. In addition, MExiCo is used indirectly by users of the Phoibos corpus manager.

CONCLUSION

The MExiCo library, and several projects related to it, are under active development. In the immediate future, MExiCo will be able to create improved RDF and metadata representations of its corpora and resources, especially in the CMDI format (Broeder, Uytvanck, Windhouwer, Gavrilidou, & Trippel, 2012). Also, there are plans for the integration of additional file formats, such as motion tracking data, and data sets with mental representations of artificial agents.

MExiCo is made available under the Gnu Lesser General Public License (LGPL) 3.0, and its releases are made available as a repository on Github (https://github.com/sfb673/mexico). The complete documented source code can be obtained from there, and alterations, improvements and comments are highly welcome.

ACKNOWLEDGEMENTS

This Research Has Been Supported By The German Research Foundation (Dfg) In The Project X1 " Multimodal Alignment Corpora: Statistical Modelling And Information Mangement"Of The Collaborative Research Centre (Crc)673 " Alignment In Communtion" Bielefeld University, Germany.

REFERENCES

1. Barras, C., Geoffrois, E., Wu, Z., & Liberman, M. (2001). Transcriber: development and use of a tool for assisting speech corpora production. *Speech Communication.*

2. Bergmann, K., & Kopp, S. (2012). Gestural Alignment in Natural Dialogue. In R. P. Cooper, D. Peebles, & N. Miyake (Eds.), *Proceedings of the 34th Annual Conference of the Cognitive Science Society (CogSci 2012)* (pp. 1326 1331). Cognitive Science Society.

3. Bird, S., & Liberman, M. (2001). A formal framework for linguistic annotation. *Speech communication, 33*(1-2), 23 60. Boersma, P., & Weenink, D. (2001). Praat, a system for doing phonetics by computer. *Glot International, 5*(9/10), 341 345.

4. Broeder, D., Uytvanck, D. Van, Windhouwer, M., Gavrilidou, M., & Trippel, T. (2012). Standardizing a Component Metadata Infrastructure. In N. Calzolari (Ed.), *Proceedings of the Eighth International Conference on Language Resources and Evaluation (LREC 2012), Istanbul* (pp. 1387 1390). European Language Resources Association (ELRA).

5. Brugman, H., & Russel, A. (2004). Annotating Multi-media / Multi-modal resources with ELAN. *Fourth International Conference on Language Resources and Evaluation.*

6. Chiarcos, C., Nordhoff, S., & Hellmann, S. (Eds.). (2012). *Linked Data in Linguistics.* Berlin, Heidelberg: Springer Berlin Heidelberg.

7. Ehlich, K., & Rehbein, J. (1976). Halbinterpretative Arbeitstranskriptionen (HIAT). *Linguistische Berichte.*

8. Evert, S., Carletta, J., O'Donnell, T. J., Kilgour, J., Vögele, A., & Voormann, H. (2003). The NITE Object Model. *Proceedings of the EACL Workshop on Language Technology and the Semantic Web* (pp. 1 17).

9. Kipp, M. (2001). Anvil   a generic annotation tool for multimodal dialogue. *Seventh European Conference on Speech Communication and Technology.*

10. Lücking, A., Bergmann, K., Hahn, F., Kopp, S., & Rieser, H. (2010). The Bielefeld Speech and Gesture Alignment Corpus (SaGA). *Proceedings of the LREC 2010 Workshop "Multimodal Corpora – Advances in Capturing, Coding and Analyzing Multimodality"* (pp. 92 98). http://www.lrec-conf.org/proceedings/lrec2010/workshops/W6.pdf

11. MacWhinney, B. (2000). *The CHILDES Project. Tools for analyzing talk.* Mahwah, NJ: Lawrence Erlbaum.

12. Menke, P., & Mehler, A. (2011). From experiments to corpora: The Ariadne Corpus Management System. *Corpus Linguistics Conference 2011.* Birmingham, UK.

13. Pickering, M. J., & Garrod, S. (2004). Toward a mechanistic psychology of dialogue. *Behavioral and brain sciences, 27*(2), 169 190.

14. Schmidt, T. (2002). Gesprächstranskription auf dem Computer das System EXMARaLDA. *Gesprächsforschung, 3,* 1 23.

15. Schmidt, T., & Schütte, W. (2010). FOLKER: An Annotation Tool for Efficient Transcription of Natural, Multi-party Interaction. In N. C. (Conference Chair), K. Choukri, B. Maegaard, J. Mariani, J. Odijk, S. Piperidis, M. Rosner, et al. (Eds.), *Proceedings of the Seventh International Conference on Language Resources and Evaluation (LREC'10).* Valletta, Malta: European Language Resources Association (ELRA).

16. Schmidt, T., & Wörner, K. (2005). Erstellen und Analysieren von Gesprächskorpora mit EXMARaLDA. *Gesprächsforschung, 6,* 171 195. Selting, M., Auer, P., & Barth-Weingarten, D. (2009). Gesprächsanalytisches Transkriptionssystem 2 (GAT 2).

17. TEI Consortium. (2008). *TEI P5: Guidelines for electronic text encoding and interchange.* TEI Consortium.

18. Van Uytvanck, D., Stehouwer, H., & Lampen, L. (2012). Semantic metadata mapping in practice : the Virtual Language Observatory. In N. Calzolari (Ed.), *Proceedings of the Eighth International Conference on Language Resources and Evaluation (LREC 2012), Istanbul* (pp. 1029 1034). European Language Resources Association (ELRA).

19. Wagner, P., Inden, B. Malisz, Z., & Wachsmuth, I. (2012). "Ja, mhm, ich verstehe dich" - Oszillator-basiertes Timing multimodaler Feedback- Signale in spontanen Dialogen. In M. Wolff (Ed.), *Elektronische Sprachsignalverarbeitung 2012 (Tagungsband ESSV). Studientexte zur Sprachkommunikation* (pp. 179 187). TUD Press.

20. Wittenburg, P., Brugman, H., Russel, A., Klassmann, A., & Sloetjes, H. (2006). Elan: a professional framework for multimodality research. *Proceedings of Language Resources and Evaluation Conference (LREC).*

CHAPTER 6

Developing library school student's research skills through assignments in a collection management course

Yelena Pancheshnikov

University of Saskatchewan, Canada, 3 Campus Drive, Saskatoon, SK ,S7N 5A4

ABSTRACT

Schools of Library and Information Science in Canada and in the U.S. are graduate professional schools that train librarians and information specialists and pursue a combination of practical and academic educational goals. Students are introduced to research through a Research Methods course, however further application of research skills is limited by the lack of a requirement for a Master's Theses. The paper discusses ways of introducing research through assignments in a Collection Management class. Assignments were designed to address the main content aspects of the discipline through the application of a research process in such standard formats as technical reports, short research proposals and presentations.

KEYWORDS

Schools of Library and Information Studies, teaching collection management, teaching research methods, assignments

1. INTRODUCTION

Library Degrees are graduate level professional degrees that are granted by over fifty U.S. and Canadian Universities. In the course of the last two decades the names of most of these schools have changed from "Library Schools" to "Schools of Library and Information Studies", or "Schools of Information Management" or "Schools of Information". Regardless to the title all of them continue to train professional librarians and grant a Master's Degree in the field of Library and/or Information Science. This degree is a required qualification for any professional librarian's position.

The need for librarians to combine practice and research in their jobs is widely recognized. To address this need the curriculum of a Library School includes a course on Research Methods, however does not include a requirement for a Master's Theses. Writing a thesis is optional. It can be substituted and in most of the cases is substituted by a portfolio of course work. Opportunities for students to develop research skills although encouraged through different venues, are limited to the content of individual courses.

Teaching research in Library Schools is also shaped by the diversity of student's educational backgrounds. The majority of students come with an undergraduate degree, although some have a Master's and a Ph.D. Most of the students have a background in the humanities, fewer – in the social sciences and very few- in the sciences and engineering. The level of previous education and the subject background inevitably affects the understanding of the research process and determines significant differences in the student's familiarity with research methodologies.

For most of the students a research methods class that is not supported by opportunities to apply the acquired knowledge is insufficient. Surveys of both -library school students and librarians conducted in the last decade show that most of the respondents consider more training and broader exposure to research very beneficial.

2. GOAL, OBJECTIVES AND MATERIAL

The purpose of the work summarized in this paper was to develop a teaching strategy that contributes to the development of research skills among library school students through their application to the content of a practical professional course. More specific question included:

-identifying priority areas of the course the study of which could be improved by a more academic research based approach

-identifying priority aspects of research in library science that can be realistically taught in the conditions of a professional course when opportunities to work on a research project comprehensively are limited- devising concrete assignments that address the two objectives identified above

The approach discussed in this paper was developed as a result of teaching a course on Collection Management. The course is offered in all library schools and although usually elective - is traditionally considered central to the profession. The content of the course covers a broad range of issues related to the development of library collections in different types of libraries and information centers and therefore opens broad opportunities to place research into a practical and professionally meaningful context.

The course was taught by the author at San Jose State University since 2008 in a fully online environment. The content was delivered in weekly modules each of which included a weekly document outlining the learning goals and objectives of learning, lectures and readings. Required activities included participation in weekly discussions and six separately graded assignments. These two types of required activities were used as venues for introducing research.

3. PROFESSIONAL ROLES OF LIBRARIANS AND THE NEED TO TEACH RESEARCH METHODS: A BRIEF REVIEW OF LITERATURE

A professional discipline like librarianship is based on two forms of knowledge: theoretical and practical. Theoretical knowledge is produced as a result of research and is disseminated primarily through scholarly channels, publications being the main of them. Practical knowledge is created as a result of everyday professional practice and is disseminated through various channels including professional publications. There obviously is a strong connection between these two forms.

The importance of the connection between research and practice for the library profession was addressed in a number of publications. Powell, Baker, and Mika (2002) indicate that research contributes to the advancement of the discipline through producing new knowledge. This new knowledge lays foundations for the improvement and innovation in professional practice and enhances the critical thinking and problem-solving ability of practitioners. Park (2003) differentiates between two levels of competencies both of which are necessary for practicing in the field of library and information science: the consumer level and the contributor level. At the consumer level library and information science practitioners read, evaluate and make use of research publications. At the research contributor level they produce research. The two levels are closely connected.

Integration of research into practice is especially important in the academic library world. The formal side of it is determined by the fact that librarians working at university libraries usually have faculty status and are required to publish for tenure and promotion. Traditional forms of library practice, such as reference consultations, instruction and various library management activities are known to have improved as a result of a more academic approach to work (Bodi, 2002). A more recent tendency in the academic library work is to encourage librarians to collaborate with faculty and graduate students through their involvement into the research process by participating in the preparation of systematic reviews, participation in grant application by aggregating and analyzing research metrics, and data curation.

In spite of the broad understanding of research as an integral part of the profession opinions were expressed about the insufficient integration of research into practice (Riggs, 1994, McClure, 1989;) Reason leading to this are numerous, but many stem out of the insufficient training received by library school students in areas pertinent to research.

A substantial number of publications discuss research skills as an essential competency of professional librarians that has to be developed through teaching research methods within the library school curriculum. These publications are based either on surveys conducted among the involved groups (library school faculty, library school students and working librarians), or on reviews of the library school curriculum and syllabi. The following brief overview provides examples of findings and thoughts supporting the need to look for ways of strengthening research training of librarians.

Stephenson (1990) reported the results of a detailed survey of instructors teaching a research methods class in over 50 accredited U.S. and Canadian Library Schools. One of her findings pertained to the very limited opportunities to practice research skills by working on concrete projects because not all of the research methods courses included a requirement for a research project.

Park (2003) reviewed the curriculum of a similar number of accredited U.S and Canadian library schools with respect to their inclusion of a research methods course. According to the findings only in half of the library schools a research methods course was offered as a core course, although at the same universities research methods classes were offered in most of the science and social science programs, and most importantly – in other graduate professional programs such as Business and Social Work.

Andersen (2002) analysed the results of a survey conducted among Master's students at the University at Albany, State University of New York. The results of the survey are discussed in the context of developing student's analytical skills in order to meet the changing requirements of the profession. According to the obtained results research was recognized by the library students as an important part of their education.

Luo (2011) conducted a detailed survey of librarians working in different types of libraries with respect to their knowledge of research methods and the usefulness of courses on research methods offered in library schools. The article describes many interesting findings and makes useful practical suggestions, such as offering continuing education opportunities and more flexibility within the curriculum.

4. RESEARCH AS A COMPONENT OF ASSIGNMENTS IN A COLLECTION MANAGEMENT COURSE

Two aspects had to be considered in the development of the assignments:

- the content part of collection management to be addressed in the assignments

- ways of integrating research into assignments in order to facilitate

a more academic approach to mastering the content of collection
management.

The choice of the core areas of collection management was based on a
traditional understanding of the scope of the discipline reflected in texts
and most importantly - on concrete projects that a collection
management librarian handles at work. Knowledge and skills necessary
for performing the most common collection management tasks
determined the choice of the following content areas for the assignments:
1.studies of user's needs 2.development of collection management policies
3.selection of library materials and 4. evaluation of library collections. A
separate assignment required an in-depth study of one of the core
collection management topics chosen by the student.

Several ways of introducing research into the content of the course were
considered. Following the system of teaching described in texts on basic
research methods in the social sciences was impossible. The limitations of
a practical course made it necessary to devise a more focused approach
tailored to the concrete content of the discipline. The decision was made
to emphasize the importance of all standard components of a research
process, but focus on the formulating of the objective and research
questions, familiarity with the basics of research design, familiarity with
methods of data collecting and analysis that are most commonly used in
library science and applicable to collection management. For this purpose
a combination of several sources was reviewed: categories related to
research in the field of library and information described by Hernon
(2001) and B.Widemuth (2009), methods of research identified by Peritz
(1980) as the most commonly used in library science publications, results
of the survey of instructors teaching a research methods class in Library
Schools reported by Stephenson (1999) and research methods described
by Eldredge (2004) for the field of medical librarianship.

Both – the collection management content and the research aspect were
introduced through two types of assignments: weekly discussions of the

studied topics and five separately graded short project types of assignments.

4.1 Discussions of a topic on the basis of articles from scholarly journals as an assignment

Learning to work with scholarly literature was addressed as an essential component in the development of student's research skills and as a foundation for the separately graded assignments that required work with original data. In many cases selected readings were studied as examples of research in collection management.

Each week students were required to contribute to the discussion of a broad collection management topic. The discussion was based on the reading of 3-4 articles selected by the instructor from peer-reviewed sources and was directed by questions asked by the instructor. The questions were formulated with the intention to facilitate a more scholarly research-based approach to a practical library science topic by drawing attention to the importance of the theoretical background in the study of the topic, clarity in the understanding of the goals, methodology, and conclusions. Selected articles were discussed with respect to their applicability in future work.

Examples of questions posted for discussion: What is the importance of the theoretical background for the study of the problem? What elements of the methodology devised by the author of the article can be applied in similar research? What conclusions of the research described in the article are most relevant to the work of a library practitioner?

4.2 Separately graded assignments

The course included a system of six separately graded assignments. The first five assignments required research. The sixth assignment was a review of the course that was not included into this study. The use of a combination of several shorter assignments rather than one large project was more suitable for the study of the diverse content of collection management. The sequence of the assignments reflected the logic of practical collection management in all types of libraries.

Table 1. Characteristics of assignments in a collection management course

Content of Collection Management addressed by the assignment	Presentation Format	Aspects of the research process emphasized by the assignment	Type of research design
Study of user's information needs	Short research proposal	Problem statement, Theoretical framework, Objectives/ research questions, Research design	Case study, study of Special populations
Collection Policies	Technical report	Description of the phenomena or setting, Research design	Comparative study, audit
Selection of Library Materials	Technical report/article	Objectives/research questions, Data collecting, Data analysis	Case study
Evaluation of Library Collections	Technical report/article	Problem statement, Objectives/ research questions Research design Data gathering, processing and analysis	Case study, Comparative study
In-Depth study of a selected topic	Presentation	Problem statement, objectives/research questions Other elements depend on the topic	Case study

5. CONCLUSIONS AND RECOMMENDATIONS

The widely recognized need to strengthen the research component of library school education can and should be implemented along different venues including the inclusion of elements of research into practical library school courses.

Assignments in courses like Collection Management are among the most obvious ways of doing it. Using a variety of assignment types and directing students to address them as short research projects would contribute to the understanding of research and provide meaningful opportunities to practice research skills. Learning to work with scholarly literature, emphasis on the understanding of the theoretical background of library research, clarity in research design and methodology should be emphasised at every opportunity.

Written and verbal comments received from students at the end of the term showed that the systematic use of scholarly literature made the course more academic and the discipline more interesting. Some students indicated that their understanding of the professional and research opportunities in library and information science have broadened and that their interest in the discipline strengthened. Some indicated that their academic writing skills improved.

A formal structured assessment would be beneficial for the understanding of the learning outcomes of the described approach and can be a subject of a follow-up study.

REFERENCES

1.   Andersen, D. L. (2002). Teaching analytic thinking: Bridging the gap between student skills and professional needs in information science.

2.   Journal of Education for Library & Information Science, 43(3), 187-196.

3.   Bodi, S. (2002). How can we bridge the gap between what we teach and what they do? Some thoughts on the place of questions in the process of research. Journal of Academic Librarianship, 28, 109–114.

4.   Dilevko, J. (2000). A new approach to teaching research methods courses in LIS programs. Journal of Education for Library & Information Science, 41(4), 307-329.

5. Eldredge, J. D. (2004). Inventory of research methods for librarianship and informatics. Journal of the Medical Library Association, 92(1), 83-90.

6. Retrieved from http://search.ebscohost.com/login.aspx?direct= true &db=llf&AN=502918603&site=ehost-live

7. Hernon, P. (2001). Components of the research process: Where do we need to focus attention? Journal of Academic Librarianship, 27(2), 81-89.

8. Luo, L. (2011). Fusing research into practice: The role of research methods education. Library & Information Science Research (07408188), 33(3), 191-201.

9. McClure, C. (1989). Increasing the usefulness of research for library managers: Propositions, issues and strategies. Library Trends, 38(2), 280– 294.

10. Park, S. (2003). Research methods as a core competency. Journal of Education for Library & Information Science, 44(1), 17-25.

11. Peritz B. (1980). The Methods of Library Science Research: Some Results From a Bibliometric Survey. Library Research 9(17), 3-83

12. Powell, R. R., Baker, L. M., & Mika, J. J. (2002). Library and information science practitioners and research. Library & Information Science Research, 24, 49–72.

13. Riggs, D. E. (1994). Losing the foundation of understanding. American Libraries, 25, 449.

14. Stephenson, M. S. (1990). Teaching research methods in library and information studies programs. Journal of Education for Library & Information Science, 31, 49-65.

CHAPTER 7

Indoor/outdoor particulate matter concentrations and microbial load in cultural heritage collections

Mihalis Lazaridis[1,*], E. Katsivela[2], ,I. Kopanakis[1], L.Raisi[2] and ,
G. Panagiaris[3]

[1] Department of Environmental Engineering, School of Environmental
Engineering, Technical University of Crete, Polytechneioupolis, 73100
Chania,Crete, Greece.
[2] Department of Environmental and Natural Resources Engineering,
Technological Educational Institute of Crete, 73133 Chania, Greece.
[3] Department of Conservation of Antiquities and Works of Art,
Technological Educational Institute of Athens, Egaleo, 12210 Athens,
Greece.

ABSTRACT

Background

Particulate matter, microorganisms in air and environmental conditions
present a potential risk to museum collections. There are also[ ] limited
studies of simultaneous measurements of airborne particles and

microorganisms inside museums and the effects of seasonal variations. Therefore, extensive indoor/outdoor measurements of particulate matter mass/number concentrations and viable, cultivable microbial load were performed in two museums and a library in Greece for a period of 2 years at selected time intervals. The culture heritage collections are located at coastal (Historical Museum of Crete in Heraklion), urban (Criminology Museum of the University of Athens) and mountainous (*Neophytos Doukas* Library in Zagori) environments and their collections consist mainly of organic materials. Measurements of inhalable particulate mass (PM_{10}, $PM_{2.5}$, PM_1) and viable, cultivable airborne microorganism concentrations (heterotrophic bacteria, cellulose metabolizing bacteria, acid producing bacteria and mesophilic fungi) in air were performed.

Results and conclusions

The indoor PM_{10} and microbial concentrations were higher than the outdoor levels showing the influence of the indoor sources, such as the presence of people and indoor activities, as well as, anthropogenic outdoor sources, and natural emissions. Elevated PM_1 particle number concentrations were also encountered in the Historical Museum of Crete in Heraklion and the Criminology Museum of the University of Athens due to the high anthropogenic emissions of the urban areas. The lowest concentrations of viable, cultivable airborne microorganisms were measured in the Historical Museum of Crete at the coastal site, which encounters also well controlled microclimatic conditions. In comparison to the other two naturally ventilated sites, the highest concentrations of viable, cultivable airborne fungi were measured in the *Neophytos Doukas* Library at the mountainous site, whereas the highest concentrations of viable, cultivable airborne heterotrophic bacteria were measured in the Criminology Museum of the University of Athens at an urban site, where mummified tissues and dry specimens are exhibited. The closed showcases of the two museums and the library could only protect the exhibits from viable, cultivable airborne fungi, but not from specific categories of bacteria. Acid producing bacteria in the Historical Museum of Crete, cellulose metabolizing bacteria in the *Neophytos Doukas* Library, and opportunistic pathogenic heterotrophic bacteria in the Criminology Museum of the University of Athens showed to be enriched inside the closed showcases.

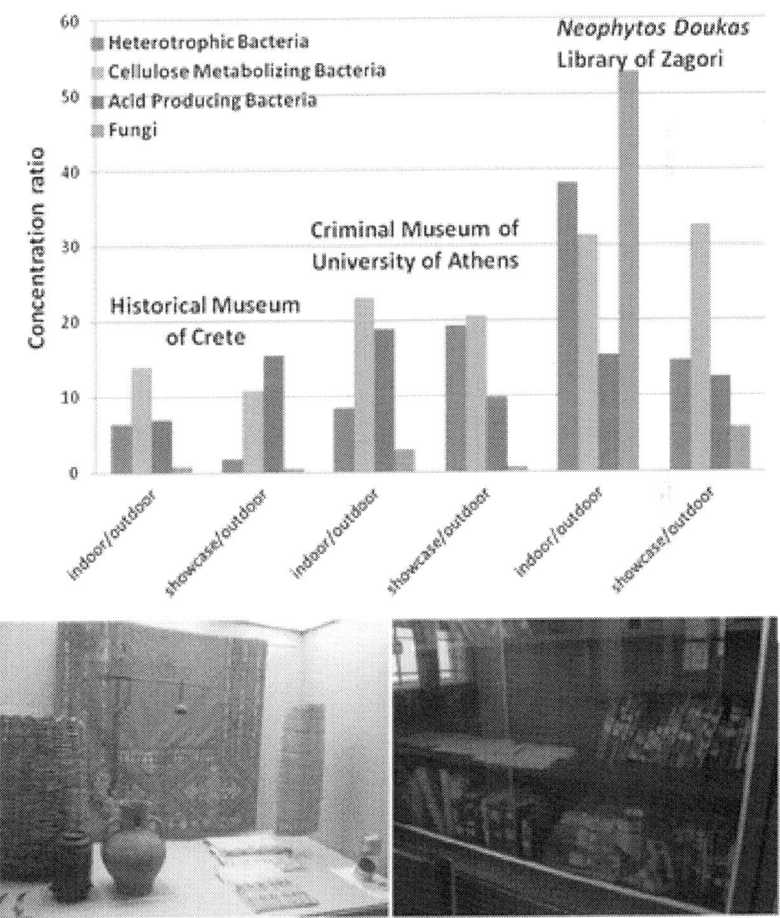

KEYWORDS

Particulate matter Bioaerosols Indoor air Museums and library

1. BACKGROUND

The indoor environment in museums is influenced by several factors, whereas the main influence arises from the environmental conditions (e.g. humidity, temperature, light), gaseous and particulate matter (PM) pollution, and microorganisms [1]. Particulate matter is a potential threat to objects in the museums due to soiling and chemical reactions from harmful compounds inside the particles or on the surface between the particle and the deposited surface. The chemical composition of particles is an important factor which affects the preservation of objects inside museums [2–7]. Furthermore, the viable, cultivable microorganism load (bacteria and fungi) in the air inside museums has been identified as a major problem for the deterioration of cultural heritage objects [8].

In particular, organic materials are very complex in structure and their deterioration is a complicated process. Besides the importance of thermal or photochemical induced oxidation process, ionic hydrolysis reactions and reactions induced by UV, there are several other factors which affect the deterioration of organic objects. Among them is humidity, temperature, particulate matter, microorganisms and gaseous pollutants (such as O_3, NO_2 or SO_2). All the reactions create changes in the organic structure caused by alterations in the chemical bonding and may lead to a disintegration of an object [1, 4].

A number of studies have been performed to determine the characteristics and sources of particulate matter and airborne microorganisms inside museums [2, 9–14].

The current study reports on particulate matter mass and number concentrations, viable, cultivable airborne microorganism concentrations, as well as, environmental conditions (infiltration rate, relative humidity, temperature) in two museums and a library in Greece for a period of 2 years. Additional measurements on PM size distribution and PM chemical composition analysis which were performed will be a subject of a future study.

2. METHODS

2.1 Sampling locations

In the current study, the measurement field studies took place in two museums and a library in Greece, with a total duration of 2 years (March 2013–February 2015). Three cultural heritage collections with different exhibits and differences in relation to location, ventilation and number of visitors were selected (Fig. 1). These cultural heritage collections are: (1) the Historical Museum of Crete (http://www.historical-museum.gr/) (coastal environment, well controlled microclimatic conditions, high number of visitors), (2) the Criminology Museum of the University of Athens (http://www.criminology-museum.uoa.gr/) (urban environment, naturally ventilated and partially air conditioned, low number of visitors), and (3) the *Neophytos Doukas* Library (mountainous environment, naturally ventilated, no visitors) at Ano Pedina, Zagori (see Fig. 1). The exhibits in the museums consist mainly of organic materials (textile, leather skin, parchment, bones, wood, paper, etc.). The measurements in each museum occurred at three sites [outdoors (site A), indoors (site B) and inside showcases (site C)] in order to determine the indoor/outdoor relationships and the effectiveness of showcases in protecting organic exhibits.

The Historical Museum of Crete is located close to the harbor of Heraklion and its distance from the sea is about 200 m. The test site is located on the second floor of the building. The Museum is housed in a neoclassical building and was founded in 1953. The building is consisted of 19 rooms and 2 court yards. The exhibits in the Museum are mainly composed of organic materials, such as paper, wood, silk, wool, cotton and paintings/panels. Site A is situated outdoor of the museum at a point behind the forefront of the building. Site B is situated inside an ethnographic exhibition room of organic materials, and the third site (site C) is inside of a showcase, which exhibits mainly wool and cotton materials. The museum has an independent mechanical ventilation system, controlled indoor conditions and a large number of visitors.

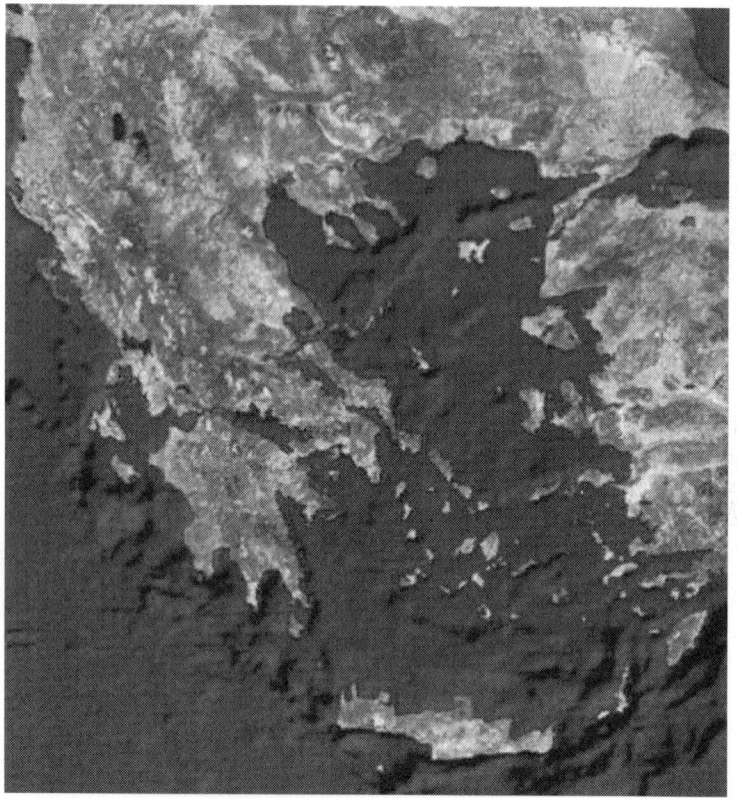

Figure 1 The three sampling locations in Greece. *a* Heraklion (Historical Museum of Crete); *b* Athens (Criminology Museum of the University of Athens); *c* Zagori (*Neophytos Doukas* Library)

The Criminology Museum of the University of Athens is situated in the centre of the town, in the region of Goudi at an altitude of 135 m. The museum is hosted in an air conditioned building, although the sampling room itself is not air conditioned. The lighting of the room is both natural and artificial, whereas the openings are partially covered with curtains. The exposed collection includes mainly organic materials, such as mummified tissues, bones, leather skin, textile, wood, paper, metal guns and knives. Measurements were performed at three sites. The first one (site A), is located outside the building, in the surroundings of the museum. The second one is located inside the museum (site B) in the centre of the exhibition room. The third site (site C) is located inside a showcase. The museum is open mainly for students from the Medical Faculty.

The *Neophytos Doukas* Library, at Ano Pedina village of Zagori region at the Prefecture of Ioannina (rural and mountainous environment) is located in a sparsely populated remote area, at an altitude of about 1000 m. The museum is hosted in the old school of the village without any air conditioning and heating system. There are four windows without curtains so that the light can penetrate unhindered indoors. The exposed collection includes mainly organic materials, such as old books most aged 2–3 centuries. The books are made of materials such as paper, leather and thread. The measurements were performed into three different locations. Site A is located outdoors, site B is located indoors and site C inside a showcase filled with old books close to a window. The museum is open to the public after appointment.

An extensive measurement campaign was conducted over a 2 years period including environmental parameters, gases, particulate matter and viable, cultivable airborne microorganisms (bioaerosol) sampling. In the current work the focus is on the particulate matter and bioaerosol measurement characteristics.

2.2 Particulate matter and bioaerosol sampling

The measurement schedule included a 2 days campaign every third month for each site to examine the seasonal variability influence for a period of 2 years.

The PM_{10}, $PM_{2.5}$, PM_1 particulate matter mass measurements in the museums were performed using the TSI's on-line Dusttrak™ Aerosol Monitor model 8520, which is a portable, battery-operated, laser-photometer that measures continuously airborne dust concentrations (TSI 2003). A P-Trak (Ultrafine Particle Counter, Model 8525, TSI) instrument was used for the determination of the number concentration characteristics. The Dust Trak and Dust Trak II instruments are based on a 90° light scattering and have a particle size range from 0.1 to 10 µm at a flow rate of 1.7 L/min with a concentration range from 1 to 100,000 µg/m³ for Dust Trak, and a flow rate of 3.0 L/min with a concentration range from 1 to 150,000 µg/m³ for Dust Trak II, respectively. The log interval was set up to 1 min. The P-Trak uses high-purity isopropyl alcohol to grow microscopic particles for easier detection and counting in the optical chamber. Its concentration range is between 0 and 5×10^5 particles/cm³ and it is able to measure particle concentration in the size range from 0.02 to 1 µm at a sample air flow rate of 0.1 L/min.

The Dusttrak™ instrument measurements were corrected based on the beta attenuation methodology. Comparative measurements of PM_{10} concentrations by the beta attenuation monitor (FH 62 SEQ) and the Dusttrak™ Aerosol Monitor model 8520 were performed at the background station for a period of 12 months. The correction formula can be expressed as: beta [Attenuation Monitor Concentration] (µg/m³) = 0.80 [Dusttrak Monitor Concentration] (µg/m³) + 10.4 [15].

In addition, the MAS 100 one stage sampler was used to collect viable particles larger than 1.1 µm with a flow rate of 100 L/min and coincides with the level 5 of the Andersen sampler. The collected air volume by the MAS 100 sampler varied from 50 to 250 L. The referred air volumes of the collected samples using the MAS 100 sampler were optimized in experiments before, so that, for reliable results, the colony number per plate (90 mm diameter agar Petri dishes) not to exceed the number of 80.

Each aluminum orifice stage was disinfected using wipes containing 70 % isopropyl alcohol between collections of different samples.

The methodology of analysis for the viable, cultivable, airborne microorganisms is based on the cultivation of the air sampled microorganisms on specific microbiological growth media. The heterotrophic bacteria were cultivated in Tryptone Soy Broth (Merck, Germany) containing 1.5 % (w/v) Agar at 37 °C in the dark for 48 h. Incubation temperature of 37 °C was chosen for the determination of the airborne, heterochthonous (not indigenous to the area of present occurrence), opportunistic pathogenic, heterotrophic bacteria due to health-related interests. The cellulose metabolizing bacteria were cultivated in Tris Phosphate Mineral Medium [16] containing 1.5 % (w/v) Agar and 0.5 % (w/v) crystalline cellulose as sole carbon source at 37 °C in the dark for 8 days. The acid producing bacteria were cultivated in the *Gluconobacter oxydans* Medium [17] at 37 °C in the dark for 8 days. Acid producing bacteria were counted only when form clear zones below the colonies. In parallel, viable mesophilic fungi were cultivated in Malt Extract Broth (Lab M, England) containing 1.5 % (w/v) Agar at 20 °C in the dark for 72 h. Air samples were collected in two sequential repetitions on each specific growth medium in three locations of each museum and library (indoors, showcase, outdoor) using a single sampler. The counted number of colonies was precisely corrected using the positive hole conversion tables, supplied by the manufactures of the used sampling instruments. Concentrations of cultivable airborne microorganisms were expressed as colony forming units per cubic meter (CFU/m^3).

2.3 Microclimate

Measurements of environmental parameters indoor/outdoor (RH, T, infiltration rate) were performed at the two museums and the library to determine the indoor microclimate and the prevailing outdoor conditions during the different seasons.

In order to measure the CO_2 levels indoors/outdoors of each museum for the determination of the air exchange rate, the IAQ instrument (Indoor Air Quality Meter, Model 8762, TSI) was used which measures the CO_2 level in parts per million (ppm). The log interval was set up to 1 min. The infiltration rate was computed by the indoor to outdoor CO_2 concentration ratio. It was found that the average infiltration rates were 1.41, 1.22 and 1.15 h^{-1} in the Historical Museum of Crete in Heraklion, the Criminology Museum of the University of Athens and the *Neophytos Doukas* Library in Zagori, respectively.

Data loggers of environmental conditions (temperature and relative humidity) were used indoors/outdoors in the museums under study. The air temperature and relative humidity were recorded continuously by three Tiny Tag data loggers in each museum placed on three locations (outdoors, indoors and in a showcase).

3. RESULTS AND DISCUSSIONS

3.1 Microclimate

The microclimate measurements showed that only at the Historical Museum of Crete the indoor conditions are well controlled since there is in operation a central air conditioning and heating system (average values: $T = 21.9 \pm 4.8\,°C$ and $RH = 54.2 \pm 4.4\,\%$ for site A, $T = 21.0 \pm 2.5\,°C$ and $RH = 51.6 \pm 4.4\,\%$ for site B, and $T = 21.3 \pm 2.4$ and $RH = 51.2 \pm 5.4\,\%$ for site C). Contrary, in the other two museums the differences between outdoor/indoor values for the relative humidity are small indicating that the indoor conditions are not well controlled. The outdoor seasonal variation of temperature gives the same profile in all three museums, with the highest values during spring and summer, and the lowest during autumn and winter. More specifically, the average temperature values at the *Neophytos Doukas* Library at Zagori showed the lowest temperatures due to its high altitude location (average values: site A: $T = 13.9 \pm 7.2\,°C$; site B: $T = 14.2 \pm 6.5\,°C$; and site C: $T = 14.2 \pm 6.6\,°C$. The relative humidity values in the *Neophytos Doukas* Library in Zagori are also close to the outdoor values indicating not well controlled conditions ($52.3 \pm 9.3\,\%$ for site A, $52.3 \pm 8.3\,\%$ for site B, $52.2 \pm 8.1\,\%$ for site C). In comparison, the average temperatures and relative humidity values at the Criminology Museum of the University of Athens were for site A $T = 20.2 \pm 6.0\,°C$ and $RH = 51.5 \pm 5.6\,\%$, for site B $20.1 \pm 5.6\,°C$ and $RH = 51.3 \pm 5.1\,\%$, and for site C $20.3 \pm 5.9\,°C$ and $RH = 50.7 \pm 5.6\,\%$.

3.2 Particulate matter concentration

Figure 2 shows the 2 years average mass concentrations of particulate matter in the different museums during the measurement campaigns using Dusttrak™ instruments. The measurements were performed for a period of 3 h at each site during the opening hours of the museums. The lowest particulate matter concentrations outdoors were observed at the *Neophytos Doukas* Library in Zagori which is located in a mountainous region at the north west of the Greek mainland, whereas the highest particle levels indoors are measured in the Criminology Museum of the University of Athens. The Athens museum is situated in a densely populated urban area (Goudi) with an extensive road network. In addition, the $PM_{2.5}$ fraction in the PM_{10} mass is ranging from 0.7 to 0.9 in the sampling sites indicating also that particles indoors have a significant outdoor origin. The only exception was observed in site C at the *Neophytos Doukas* Library in Zagori with a value of 0.3 as a result of an individual measurement with extreme high concentration of PM_{10} ($307\ \mu g/m^3$) at the first measurement campaign. This result can be caused by coarse mineral and organic airborne dust found in the showcases. Furthermore, the average PM_{10} mass concentration indoors is higher than the outdoor levels for the museums with natural ventilation (Athens and Zagori) and close to the outdoor levels in the well controlled Historical Museum of Crete. This finding indicates that the presence of visitors (as will be shown also in Fig. 3) and indoor activities in the museums, as well as infiltration from the outdoor atmosphere (as will be shown also in Fig. 4) contribute to higher concentrations through particle resuspension and penetration. In addition, the well controlled microclimate conditions and the clean environment in the Historical Museum of Crete in Heraklion result that the indoor levels are not considerably elevated due to the presence of visitors (60–250 visitors per day).

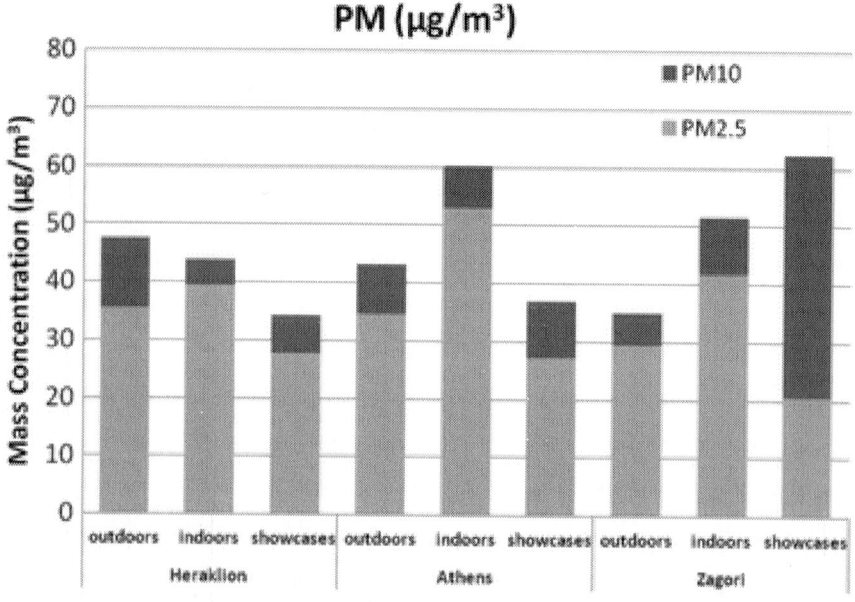

Figure 2 Two years average mass concentrations ($\mu g/m^3$) of the respirable particulate matter concentrations ($PM_{2.5}$, PM_{10}) in the two museums and the library

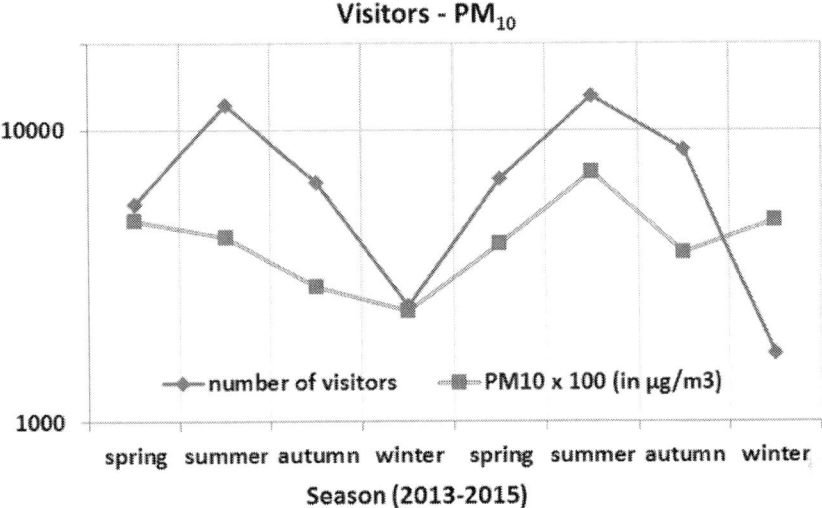

Figure 3 Seasonal variation of the number of visitors in the Historical Museum of Crete versus the seasonal average indoor PM_{10} mass concentrations ($PM_{10} \times 100$ µg/m³)

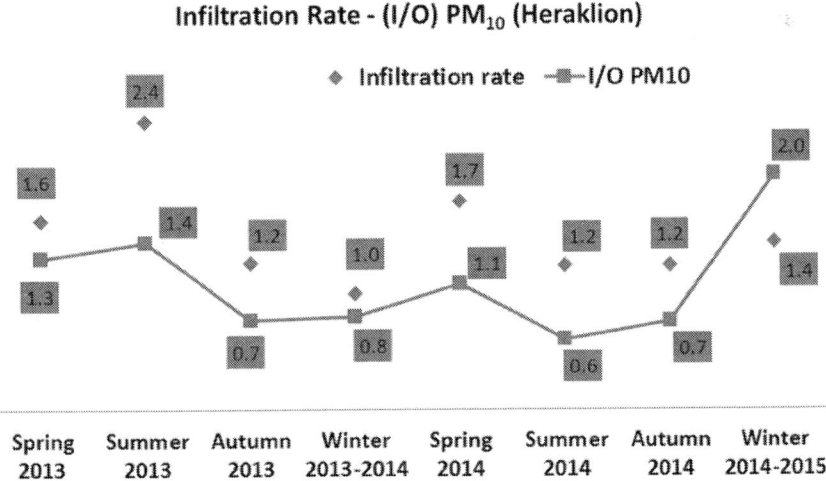

Figure 4 Seasonal variation of the infiltration rates in the Historical Museum of Crete in conjunction with the Indoor/Outdoor (I/O) PM_{10} mass concentrations

In particular, the average PM_{10} concentrations outdoors (site A) were $48.4 \pm 25.5\ \mu g/m^3$ in Heraklion, $45.5 \pm 12.3\ \mu g/m^3$ in Athens and, $37.8 \pm 14.9\ \mu g/m^3$ in Zagori for the whole time period of 2 years, as shown in Fig. 2. Furthermore, the PM_{10} mass concentration indoors was higher than the outdoor levels in the Criminology Museum of the University of Athens and the *Neophytos Doukas* Library in Zagori respectively, which indicates indoor sources such as particle resuspension due to the presence of people inside the museums and poor cleaning conditions. In addition, the indoor concentration in the Historical Museum of Crete in Heraklion was close to the outdoor levels as shown in Fig. 2, which indicates the influence of outdoor PM concentrations (as shown also in Fig. 4).

A seasonal variation of the PM_{10} and $PM_{2.5}$ concentrations is shown in Fig. 5. The highest indoor concentrations recorded during the autumn and winter periods in the Criminology Museum of the University of Athens (76.5 and $70.5\ \mu g/m^3$, respectively), whereas comparable high was the average PM concentration in autumn inside the showcase in the *Neophytos Doukas* Library in Zagori. However, the lowest average concentrations indoors were recorded in the *Neophytos Doukas* Library in Zagori varying from $28.5\ \mu g/m^3$ in spring to $47.5\ \mu g/m^3$ in summer. As shown in Fig. 5, the highest concentrations of particulate matter in the Historical Museum of Crete in Heraklion were recorded during summer and spring (57.5 and $45.0\ \mu g/m^3$, respectively).

The PM_{10} concentrations outdoors were lower than the Air Quality Standard annual level of $40\ \mu g/m^3$ (AQS annual level). However, the outdoor concentrations of the PM_{10} in Athens during the autumn, in Heraklion during autumn and summer and in Zagori during summer were higher than $40\ \mu g/m^3$ (Fig. 5). The elevated PM_{10} levels can be attributed to an increased vehicular traffic and other anthropogenic sources in the Athens and Heraklion metropolitan area in agreement with previous studies [18]. In addition, during summer there is an increased traffic in Zagori area from tourists.

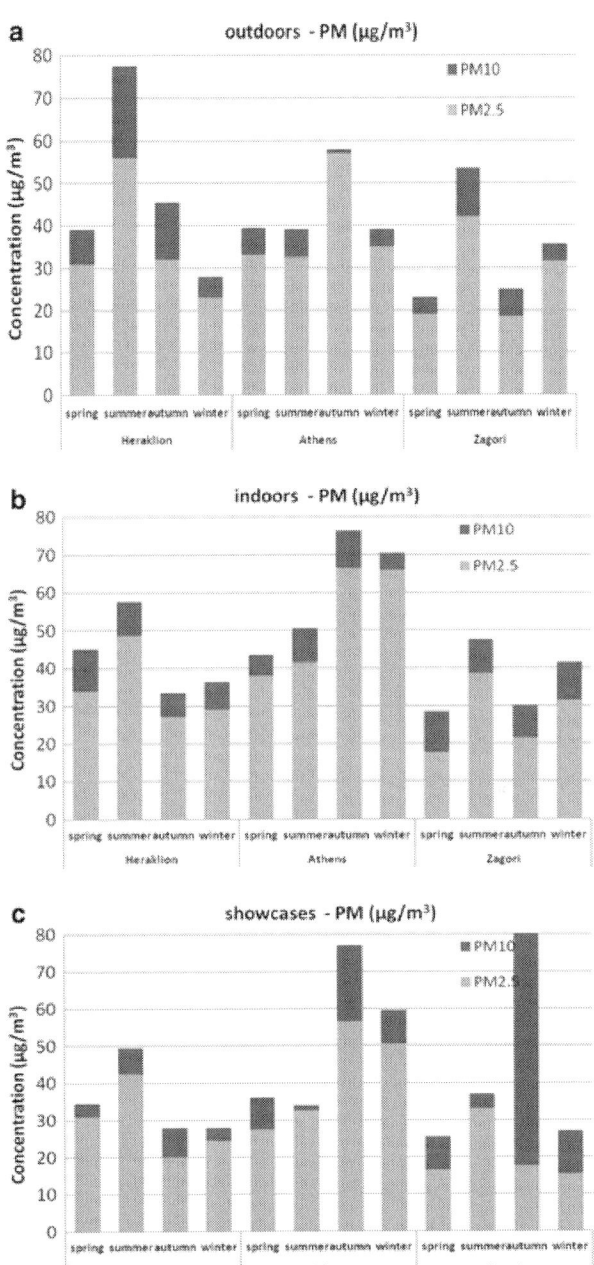

Figure 5 Seasonal variation of the $PM_{2.5}$ and PM_{10} (µg/m³) concentration in the Historical Museum of Crete in Heraklion, the Criminology Museum of the University of Athens and the *Neophytos Doukas* Library in Zagori during the measurement 2 year period (March 2013–February 2015) at **a** outdoors; **b** indoors; and **c** showcases)

As mentioned before, the indoor PM_{10} concentrations were higher than outdoor levels nearly all seasons in all cultural heritage collections due to the presence of people, ventilation conditions and activities indoors. Furthermore, the concentration of PM_{10} and $PM_{2.5}$ in showcases shows often similar values with the indoor levels indicating that the use of showcases is not always an effective mechanism for reducing the exposure to particles. Similar results are described also in the section *Bioaerosol measurements* for the dual role of the showcases in the museums (protective although in some cases result to an enrichment of microbial threats) regarding the protection of the exhibits from viable, cultivable airborne microorganisms with possible degradation capabilities. In addition, higher indoor PM_{10} concentrations than the AQS level may have an impact also to the soiling risk of exhibits in museums and air quality data may further used for soiling risk studies.

The indoor $PM_{10}/PM_{2.5}$ ratio for the three cultural heritage collections is shown in Fig. 6 for indoor, showcase and outdoor conditions. The highest indoor $PM_{2.5}/PM_{10}$ ratio is observed in the Criminology Museum of the University of Athens (0.88) and the lowest in the *Neophytos Doukas* Library in Zagori (0.64). This means that in the case of the Criminology Museum of Athens indoors, fine particles dominate. High $PM_{2.5}/PM_{10}$ ratios are also observed outdoors which indicate that fine particles are originating outdoors from anthropogenic sources. At the same time higher than one values of the Indoor/Outdoor (I/O) PM_{10} concentration ratio (as in the case of the two naturally ventilated collections in Athens and Zagori, Table 1) indicate significant indoor sources. Furthermore, the lower $PM_{2.5}/PM_{10}$ ratio at the *Neophytos Doukas* Library in Zagori shows the presence of coarse particles mainly from soil resuspension processes. The importance of particle resuspension on PM_{10} mass concentration indoors has been studied before [19] concluding that human walking and wind flow can increase considerably the PM_{10} mass concentration levels.

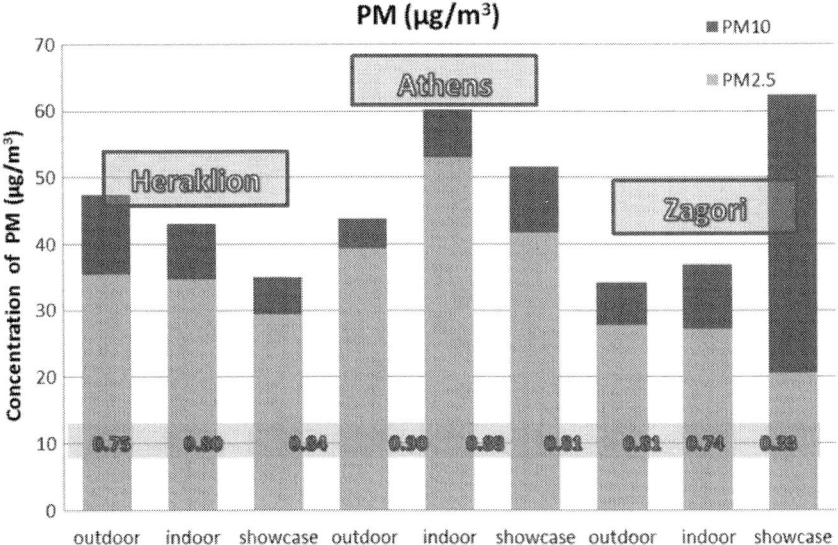

Figure 6 Indoor $PM_{2.5}$ and PM_{10} concentrations and $PM_{10}/PM_{2.5}$ ratios (indicated with *decimal numbers on the bars* of the figure) for the Historical Museum of Crete in Heraklion, the Criminology Museum of the University of Athens and the *Neophytos Doukas* Library in Zagori in the different measurement sites

Table 1 Seasonal variation of the 2 years average $PM_{2.5}/PM_{10}$ ratios indoors and the indoor/outdoor (I/O) PM_{10} ratios.

Museum	Season	$PM_{2.5}/PM_{10}$	Mean value	I/O (PM_{10})	Mean value
Historical Museum of Crete	Spring	0.76	0.80	1.15	0.98
	Summer	0.84		0.74	
	Autumn	0.81		0.74	
	Winter	0.79		1.30	
Criminology Museum of the University of Athens	Spring	0.87	0.88	1.10	1.38
	Summer	0.82		1.29	
	Autumn	0.87		1.32	
	Winter	0.94		1.81	
Neophytos Doukas Library in Zagori	Spring	0.61	0.73	1.24	1.12
	Summer	0.81		0.89	
	Autumn	0.72		1.20	
	Winter	0.76		1.17	

The seasonal variation of the $PM_{10}/PM_{2.5}$ ratio indoors for the three cultural heritage collections is shown in Table 1. Elevated I/O PM_{10} values are measured mainly in the Criminology Museum of the

University of Athens and the *Neophytos Doukas* Library in Zagori (2 years average I/O PM_{10} values 1.38 and 1.12, respectively), which are correlated with high dust load in the collections and resuspension of particles during the presence of people. The Historical Museum of Crete in Heraklion had a mean value of the I/O PM_{10} ratio close to one (0.98) indicating that there are no significant indoor particle sources.

Furthermore, Fig. 4 presents the mean seasonal variation of the infiltration rate of the Historical Museum of Crete in Heraklion versus the I/O PM_{10}. There is a correlation between the infiltration rates with the I/O PM_{10} values which indicates the influence of the outdoor particle levels in the indoor PM_{10} concentrations. Higher infiltration rates result to higher I/O PM_{10} values and therefore to an increased indoor PM_{10} concentrations.

Moreover, the influence of the number of visitors to the indoor mean PM_{10} concentrations is studied for the Historical Museum of Crete in Heraklion where detailed data are available for the number of visitors as depicted in Fig. 3. The Pearson coefficient for the 2 years measurements is 0.76 which supports the hypothesis of the influence of visitors to the indoor PM_{10} levels through particle resuspension [19] and the visitors' presence [13]. Similarly, the concentrations of the cultivable, airborne heterotrophic bacteria correlated well with the daily visitors numbers showing a Pearson coefficient for the 2 years measurements of 0.75, indicating their anthropogenic origin. In contrast, the cultivable airborne fungi did not show any correlation with the number of visitors, probably, due to their outdoor origin. These results coincide well with the Indoor/Outdoor ratios of the viable, cultivable airborne heterotrophic bacteria and fungi, which are presented and discussed in the section *Bioaerosol measurements*.

Figure 7 presents the PM_1 particle number concentration (particles/cm^3) in the three cultural heritage collections for the 2 years measurement period. The mean number concentration indoors in the Historical Museum of Crete in Heraklion and the Criminology Museum of the University of Athens show higher concentrations (mean value of 10,000 particles/cm^3) in all three sites compared to the *Neophytos Doukas* Library (mean value of 2000 particles/cm^3). The high particle number concentrations are due to the high anthropogenic emissions in the urban areas of Athens and Heraklion compared to the Zagori, which is located in a mountainous region. Similar results were observed in the Baroque Library Hall in Prague, which is also located in an urban setting [12]. The indoor number particle concentrations is higher compared to the outdoor levels in both Athens and Heraklion museums for all seasons except the winter period in Athens metropolitan area indicating indoor sources. The higher levels of outdoor particle number concentration during the winter period are due to heating and elevated vehicular traffic. The higher indoor PM$_1$ particle concentrations are encountered mainly during winter, summer and autumn periods which are associated with an increased number of visitors in the Historical Museum of Crete and the use of the museum cafeteria, as well as, due to elevated outdoor particle levels. Furthermore, the lowest PM$_1$ number concentration values were measured in the *Neophytos Doukas* Library (mean concentrations of particles 2036 ± 465 particles/cm^3 outdoors, 1598 ± 362 particles/cm^3 indoors, and 1545 ± 428 particles/cm^3 in showcases).

The higher PM$_1$ number concentration outdoors was encountered during winter. No significant seasonal trend is observed for the other seasons in the outdoor particle concentrations for Historical Museum of Crete and *Neophytos Doukas* Library, whereas for the Criminology Museum of the University of Athens lower values are observed in the summer associated mainly with a vehicular traffic decrease due to summer holiday period and reduction of the anthropogenic activities. In addition, the particle number concentration inside the showcases shows comparable values with the indoor concentrations indicating that showcases do not present an effective means for reducing the particle number concentration.

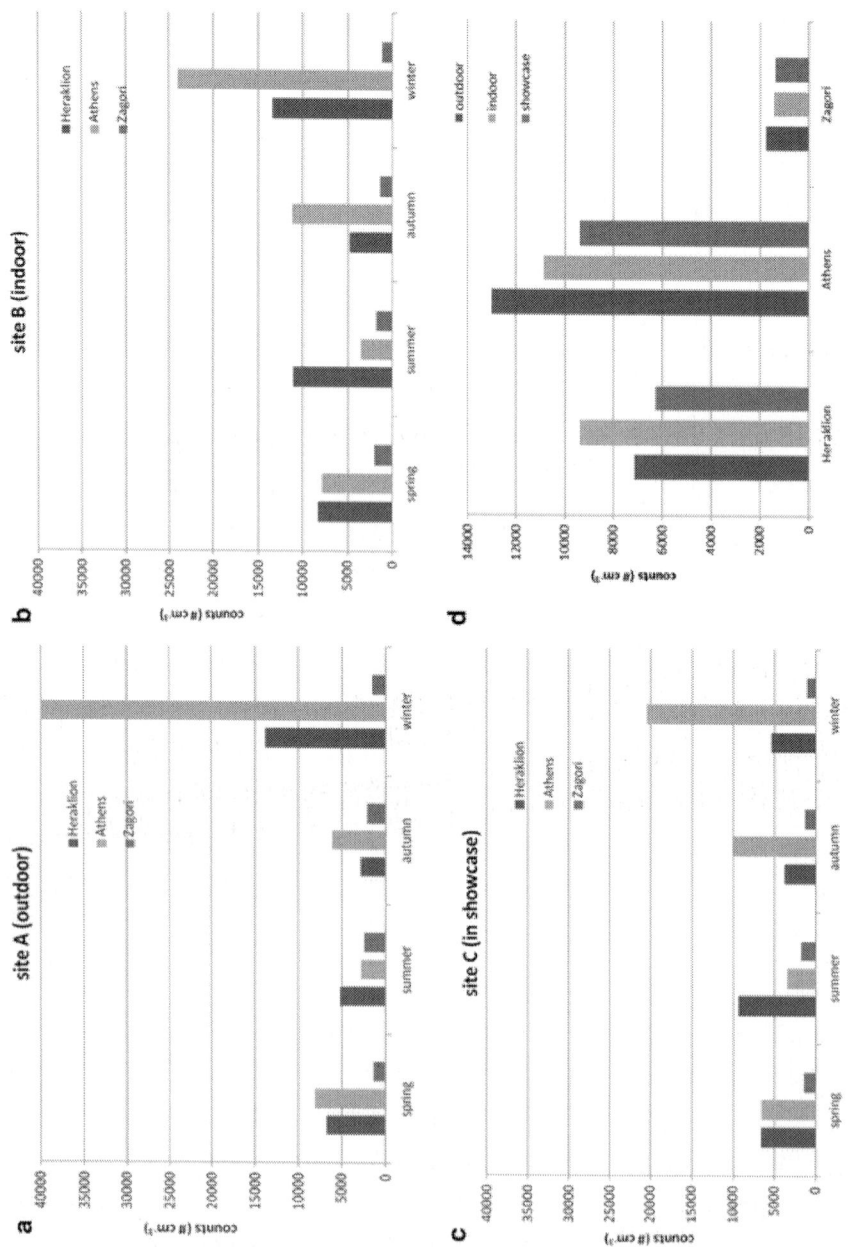

Figure 7 Seasonal variation of particle number concentration (particles/cm³) in the Historical Museum of Crete in Heraklion, the Criminology Museum of the University of Athens and the *Neophytos Doukas* Library in Zagori **a** outdoors; **b** indoors; **c** showcases and **d** mean particle number concentration

3.3 Bioaerosol measurements

Figure 8 presents the measured seasonal variation of viable, cultivable airborne microbes in the three cultural heritage collections during the 2 years measurement period. As shown, high seasonal variability of airborne microbial concentrations was measured. The lowest concentrations of cultivable, airborne microorganisms were determined in the Historical Museum of Crete (average 2 years sum indoors 2292 ± 795 CFU/m^3; showcase 1677 ± 834 CFU/m^3; outdoors 690 ± 379 CFU/m^3), whereas the highest concentrations were measured in the *Neophytos Doukas* Library of Zagori (average 2 years sum indoors $7310 \pm 4,575$ CFU/m^3; showcase 4168 ± 1201 CFU/m^3; outdoors 1266 ± 954 CFU/m^3). The cultivable bioaerosol concentrations showed the highest values in autumn indoors (average 2 years autumn sum indoors varied from 3293 to 5093 CFU/m^3). In comparison, the lowest concentrations of cultivable airborne bacteria were observed in winter (average 2 years winter sum indoors varied from 680 to 1375 CFU/m^3). Extremely high concentrations of cultivable, airborne fungi were measured during winter inside the *Neophytos Doukas* Library of Zagori (average 2 years winter concentration of airborne fungi indoors 13,283 CFU/m^3), where the microclimate was not controlled,and the average winter relative humidity varied from 65.5 to 70.3 % and the average winter temperature from 2.2 to 13.9 °C indoors.

The 2 years average indoor/outdoor and showcase/outdoor ratios of the airborne microbes is shown in Fig. 9. As reported before [20], the I/O bioaerosol concentration ratio is an important parameter for the evaluation of biological pollution indoors. I/O ratios between 1.5 and 2 indicate indoor environment of regular conditions, whereas I/O ratios higher 2 indicate not proper indoor environmental conditions [21]. As shown in Figs. 8 and 9, the indoor concentrations of the cultivable airborne bacteria were higher than outdoors. In contrast, relatively low concentrations of airborne fungi were measured indoors and inside the showcases. Exception was the high indoor fungal concentration in the naturally ventilated *Neophytos Doukas* Library of Zagori during the winter period.

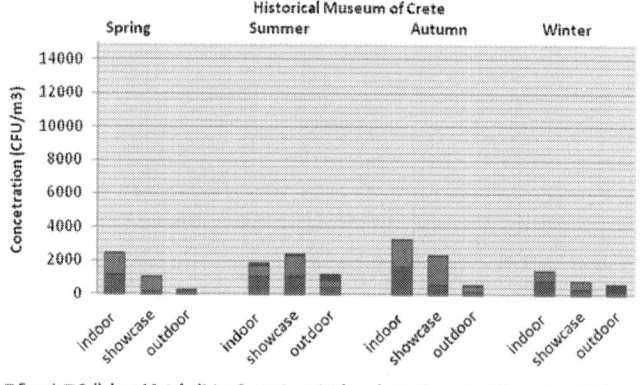

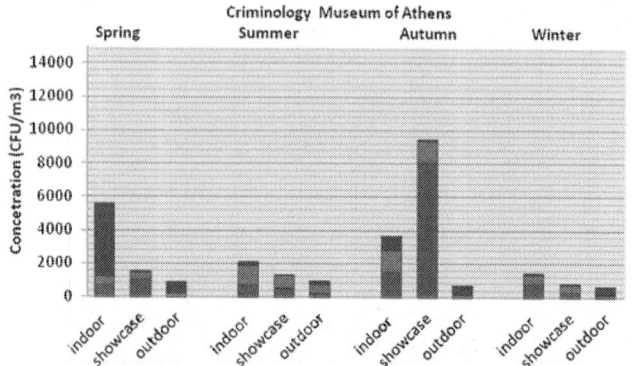

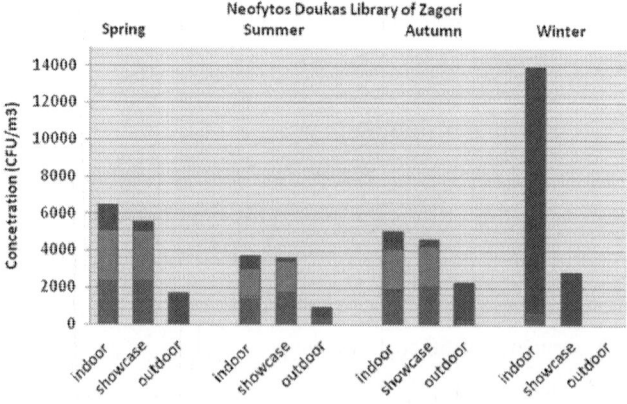

Figure 8 Seasonal variations of viable, cultivable airborne microbes in the Historical Museum of Crete in Heraklion, the Criminology Museum of the University of Athens and the *Neophytos Doukas* Library in Zagori (indoors, showcases, and outdoors). Number of samples (N = 16)

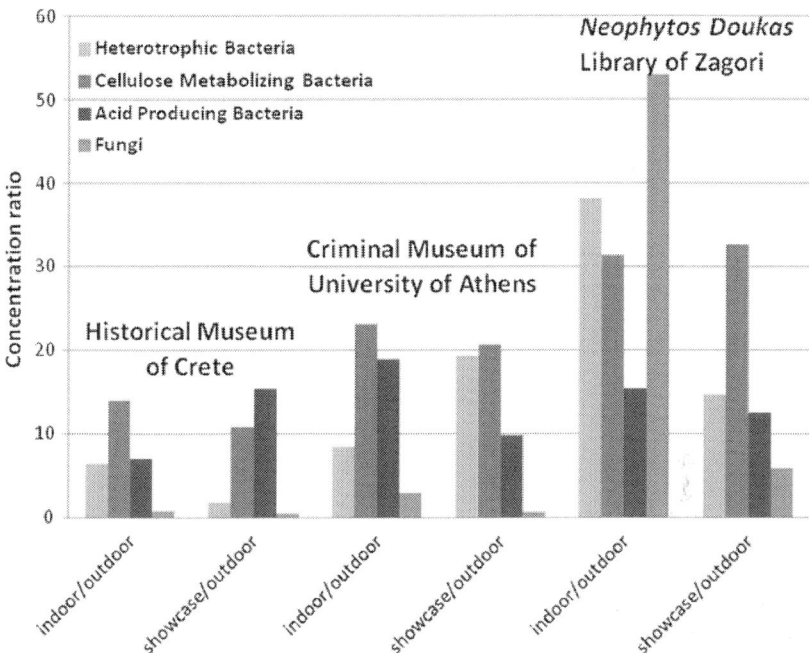

Figure 9 Two years average indoor/outdoor and showcase/outdoor ratios of the viable, cultivable airborne microbes (heterotrophic bacteria, cellulose metabolizing bacteria, acid producing bacteria, and fungi) in the Historical Museum of Crete in Heraklion, the Criminology Museum of the University of Athens and the *Neophytos Doukas* Library in Zagori. Number of samples (N = 16)

As regards the airborne fungi, the I/O and the showcase/outdoor ratios in the Historical Museum of Crete as well as the showcase/outdoor ratio in the Criminology Museum were under 1 (0.5–0.7), indicating very good storage conditions. Most of the airborne fungi come from the outdoor environment. In contrast, the 2 years average I/O ratio and the showcase/outdoor ratio in the *Neophytos Doukas* Library of Zagori were extremely high (53 and 5.8, respectively), indicating wall damping, stagnation of water and poor cleaning conditions.

Regarding the airborne bacteria, the 2 years average I/O ratios in all three collections were very high varying from 6.4 to 38.1, signalizing the enrichment of opportunistic pathogenic heterotrophic bacteria, cellulose metabolizing bacteria, and acid producing bacteria inside the two museums and the library. The most abundant bacteria in all three collections were the cellulose metabolizing bacteria, followed by the heterotrophic bacteria. The highest concentrations of the cellulose metabolizing bacteria were measured in the *Neophytos Doukas* Library of Zagori, where damages of the exhibited books are visible.

An enrichment of airborne bacteria inside the showcases was also observed indicating that showcases offer not enough protection for specific cultivable, airborne bacteria and fungi (compare in Fig. 9 the enrichment of heterotrophic bacteria during autumn in the closed showcase in the Criminology Museum of the University of Athens). Similarly to PM, showcases can only partly protect the artworks from high concentrations of airborne microorganisms with possible degradation capabilities and acid production. Especially, measurements showed that the 2 years average showcase/outdoor ratio of the acid producing bacteria was 2.2 times higher than the I/O in the Historical Museum of Crete. Similarly, the showcase/outdoor ratio of opportunistic pathogenic heterotrophic bacteria was 2.3 times higher than the I/O ratio in the Criminology Museum of the University of Athens, whereas the showcase/outdoor ratio of the cellulose metabolizing bacteria was in the same range like the I/O ratio of the *Neophytos Doukas* Library in Zagori (Fig. 9). These findings lead to the suggestion that the elevated concentrations of the museum-specific, cultivable, airborne bacteria that were enriched inside the showcases, probably, can utilize the exhibits' materials as sources for growth.

It must be mentioned, that the choice of specific selective or non-selective microbiological growth medium, the incubation temperature and the cultivation time result to differences on the selection of the viable, cultivable, airborne microbial community. Furthermore, the exact microbial concentration cannot be determined using only cultivation dependent methods since microbes may be viable but non-cultivable and culturing usually leads to underestimation of microbial diversity and concentration [22–26]. Nevertheless, the combination of culture-based analysis with molecular analysis increases the observed bacterial diversity.

In a former study [26], we have shown that PM_{10} mass and PM_1 number concentrations, relative humidity and UV radiation play a significant role in the prediction of the viable, cultivable fungal concentration in the outdoor air under specific limitations at the coastal Mediterranean city of Chania in Crete. In addition, recently Skóra et al. [27] have found a correlation between microclimatic conditions (temperature and relative humidity) and concentrations of cultivable, airborne heterotrophic bacteria and fungi in museums, archives and libraries, where the types and levels of cultivable airborne microbes were institution depended. Similarly, in the present study, we have measured "collection specific" enrichment of cultivable, airborne microbes indoors and inside the showcases. However, no significant correlation between the cultivable airborne microbes, the microclimatic conditions and the PM concentrations was found indoors or inside the showcases (Pearson coefficient less than 0.6). Exceptionally, a moderate correlation between the cultivable airborne heterotrophic bacteria and the PM_1 number concentrations was found indoors and outdoors only in the *Neophytos Doukas* Library in Zagori with Pearson *r* coefficient 0.82 and 0.83, respectively. The non significant correlation among microorganisms and PM may be explained by the fact that the PM and the airborne microbial concentrations showed different seasonal variability both indoors and outdoors in the two museums and the library.

Finally, as shown in Fig. 9, the best conditions were observed in the Historical Museum of Crete due to the regulation of the museum microclimate and the proper controlled storage conditions [28], although this museum receives a large number of visitors. In contrast, the worst conditions were measured in natural ventilated *Neophytos Doukas Library* in Zagori. From the measurements performed in the current study it can be concluded that in respect to the chemical composition and conservation of the exhibits it is possible that the indoor environment of the museums and of the showcases can result in the enrichment of microorganisms which can attack and partly deteriorate the exhibits.

4. CONCLUSIONS

Well controlled microclimatic conditions were measured in the Historical Museum of Crete. In contrast, was no significant differentiation of the indoor, and showcases microclimate as well as outdoor conditions in the natural ventilated Criminology Museum of the University of Athens and the *Neophytos Doukas* Library in Zagori.

The particulate matter measurements in the two museums and a library showed a considerable variability which is related to the outdoor concentration levels, indoor environmental conditions, infiltration rates and indoor activities such as the number of visitors. The lowest 2 years average particulate matter concentrations were observed at the *Neophytos Doukas* Library in Zagori, which is located in a mountainous region at the north west of the Greek mainland, whereas the highest particle levels are observed in the Criminology Museum of the University of Athens.

The 2 years average $PM_{2.5}$ fraction in the PM_{10} mass is ranging from 0.9 to 0.74 in the indoor and outdoor sampling sites indicating also that particles indoors have a significant outdoor origin. The average PM_{10} mass concentration and the PM_1 number concentration indoors were higher than the outdoor levels for all three museums. This finding indicates that the presence of visitors in the museums and other indoor activities contribute to higher PM_{10} mass concentrations also through particle resuspension. In addition, high PM_1 particle number concentration were measured in the Historical Museum of Crete and the Criminology Museum in Athens which are located in urban areas due to the high anthropogenic emissions.

High seasonal variability of the concentrations of cultivable, airborne microorganisms was measured in the museums. In comparison, the lowest concentrations of cultivable, airborne microorganisms were measured in the Historical Museum of Crete in Heraklion. In contrast, the highest concentrations of airborne fungi were measured in the *Neophytos Doukas* library in Zagori, whereas the highest concentrations of airborne, opportunistic pathogenic, heterotrophic bacteria were measured in the Criminology Museum of the University of Athens. Occasionally, extremely high concentrations of fungi (during winter indoors in the *Neophytos Doukas* Library in Zagori) and heterotrophic bacteria (during autumn inside the showcases in the Criminology Museum of the University of Athens) were encountered.

Although, the particulate matter correlate well with number of visitors, did not show any significant correlation with the cultivable airborne microbes, and presented a different seasonal variability versus microorganism both indoors and outdoors.

Showcases offer not always a protection for PM and specific airborne bacteria. In contrast, an enrichment of viable, cultivable airborne bacteria inside the showcases was observed in some cases that may be related to a possible utilization of the exhibits materials as sources for growth. On the other hand, although the closed showcases of the two museums could only protect the exhibits to a low percentage (8.7–9.4 %) from airborne bacteria and fungi, the concentrations of airborne microbes in the closed showcases of the *Neophytos Doukas* Library were 56.3 % reduced, indicating the importance of closed showcases, especially in museums with no microclimatic management.

An appropriate management of the museums microclimate according the ISO 11799, 2003 [28], in accordance to proper cleaning, regular dusting, disinfecting, and disinfestating of the rooms, is the best practice to protect the exhibits, and the health of people who consult them or work in them.

AUTHORS' CONTRIBUTIONS

ML conceived the study for particulate matter measurements and made substantial contribution to conception and manuscript writing. EK conceived the study for microorganisms in air and made substantial contribution to the management of the measurement campaigns and manuscript writing. IK performed the field work for particulate matter and environmental parameters. LR performed the field work for the microorganisms in air. GP managed the study and made substantial contributions to study design. All authors read and approved the manuscript.

ACKNOWLEDGEMENTS

This research has been co-financed by the European Union (European Social Fund—ESF) and Greek National Funds through the Operational Program "Education and Lifelong Learning" of the National Strategic Reference Framework (NSRF)—Research Funding Program THALES (project INVENVORG contract No. MIS 376986).

COMPETING INTERESTS

The authors declare that they have no competing interests.

REFERENCES

1. Camuffo D. Microclimate for cultural heritage. Amsterdam: Elsevier; 1998.

2. Gysels K, Delalieux F, Deutsch F, Van Grieken R, Camuffo D, Bernardi A, Sturaro G, Busse HJ, Wiser M. Indoor environment and conservation in the Royal Museum of Fine Arts, Antwerp, Belgium. J Cult Herit. 2004;5:221–30.

3. Brimblecombe P. Review article: the composition of museum atmospheres. Atmos Environ. 1990;24B:1–8.

4. Anaf W, Horemans B, Madeira TI, Carvalho ML, De Wael K, Van Grieken R. Effects of a constructional intervention on airborne and deposited particulate matter in the portuguese national tile museum, Lisbon. Environ Sci Pollut Res. 2013;20:1849–57.

5. Lopez-Aparicio S, Smolik J, Maskova L, Souckova M, Grontoft T, Ondrackova L, Stankiewicz J. Relationship of indoor and outdoor air pollutants in a naturally ventilated historical building envelope. Build Environ. 2011;46:1460–8.

6. Grau-Bove J, Strlic M. Fine particulate matter in indoor cultural heritage: a literature review. Herit Sci. 2013;1:8.

7. Smolík J, L. Mašková N, Zíková L, Ondráčková L, Ondráček J. Deposition of suspended fine particulate matter in a library. Herit Sci. 2013;1:7.

8. Brimblecombe P, Blades N, Camuffo D, Sturaro G, Valentino A, Gysels K, Van Grieken R, Busse H-J, Kim O, Ulrych U, Wieser M. The indoor environment of a modern museum building, the Sainsburt Centre for Visual Arts, Norwich, UK. Indoor Air. 1999;9:146–64.

9. Gauzere C, Moletta-Denat M, Blanquart H, Ferreira S, Moularat S, Godon J-J, Robine E. Stability of airborne microbes in the Louvre Museum over time. Indoor Air. 2014;24:29–40.

10. Gauzere C, Moletta-Denat M, Bousta F, Moularat S, Orial G, Ritoux S, Godon JJ, Robine E. Reliable procedure for molecular analysis of airborne microflora in three indoor environments: an office and two different museum contexts. CLEAN Soil Air Water. 2013;41:226–34.

11. Worobiec A, Samek L, Krata A, Van Meel K, Krupinska B, Stefaniak EA, Karaszkiewicz P, Van Grieken R. Transport and deposition of airborne pollutants in exhibition areas located in historic buildings-study in Wawel Castle Museum in Cracow, Poland. J Cult Herit. 2010;11:354–9.

12. Chatoutsidou SE, Mašková L, Ondráčková L, Ondráček J, Lazaridis M, Smolík J. Modeling of the aerosol infiltration characteristics in a cultural heritage building: The Baroque Library Hall in Prague. Build Environ. 2015;89:253–63.

13. Yoon YH, Brimblecombe P. Clothing as a source of fibres within museums. J Cult Herit. 2000;1:445–54.View Article

14. Yoon YH, Brimblecombe P. The distribution of soiling by coarse particulate matter in the museum environment. Indoor Air. 2001;11:232–40.

15. Chalvatzaki E, Kopanakis I, Kontaksakis M, Glytsos T, Kalogerakis N, Lazaridis M. Measurements of particulate matter concentrations at a landfill site (Crete, Greece). Waste Manag. 2010;11:2058–64.

16. Leibniz Institute DSMZ—German Collection of Microorganisms and Cell Cultures, 2012. 457. Mineral Medium (Brunner) [http://www.dsmz.de/microorganisms/medium/pdf/DSMZ_Medium457.pdf].

17. Leibniz Institute DSMZ—German Collection of Microorganisms and Cell Cultures, 2007. 105. Gluconobacter oxydans Medium [http://www.dsmz.de/microorganisms/medium/pdf/DSMZ_Medium105.pdf].

18. Diapouli E, Eleftheriadis K, Karanasiou AA, Vratolis S, Colbeck I, Lazaridis M. Indoor and outdoor particle number and mass concentrations in three houses in Athens. Sources, sinks and variability of aerosol parameters. Aerosol Air Qual Res. 2012;11:632–42.

19. Serfozo N, Chatoutsidou SE, Lazaridis M. The effect of particle resuspension during walking activity to PM10 mass and number concentrations in an indoor microenvironment. Build Environ. 2014;82:180–9.

20. Madrioli P, Caneva G, Sabbioni C. Cultural Heritage and aerobiology—methods and measurement techniques for biodeterioration monitoring. Berlin: Springer Verlag; 2004.

21. Ross C, de Menezes JR, Svidzinski TIE, Albino U, Andrade G. Studies on fungal and bacterial population of airconditioned environments. Braz Arch Biol Technol. 2004;47(5):827–35.

22. Amann RI, Ludwig W, Schleifer KH. Phylogenetic identification and in situ detection of individual microbial cells without cultivation. Microbiol Rev. 1995;59(1):143–69.

23. Després VR, Huffman JA, Burrows SM, Hoose C, Safatov AS, Buryak G, Fröhlich-Nowoisky J, Elbert W, Andreae MO, Pöschl U, Jaenicke R. Primary biological aerosol particles in the atmosphere: a review. Tellus B. 2012;64:15598. doi:10.3402/tellusb.v64i0.

24. Fierer N, Liu Z, Rodríguez-Hernández M, Knight R, Henn M, Hernandez MT. Short-term temporal variability in airborne bacterial and fungal populations. Appl Environ Microbiol. 2008;74(1):200–7.

25. Rappé MS, Giovannoni SJ. The uncultured microbial majority. Ann Rev Microbiol. 2003;57:369–94.

26. Raisi L, Aleksandropoulou V, Lazaridis M, Katsivela E. Size distribution of viable, cultivable, airborne microbes and their relationship to particulate matter concentrations and meteorological conditions in a Mediterranean site. Aerobiologia. 2013;29(2):233–48.

27. Skóra J, Gutarowska B, Pielech-Przybylska K, Stępień L, Pietrzak K, Piotrowska M, Pietrowski P. Assessment of microbiological contamination in the work environments of museums, archives and libraries. Aerobiologia. 2015;31:389–401.

28. ISO 11799:2003 Information and documentation—documents to rage requirements for archive and library materials.

CHAPTER 8

An Integrated Management System for Multimedia Digital Library

N. Barbuti[a]*, **S. Ferilli[b]**, **D. Redavid[c]**, **T. Caldarola[d]**

[a]*Dept. of Classical and Late Antiquity Studies - University of Bari, Piazza Umberto I, 1 70121 Bari, Italy*

[b]*Dept. of Computer Science - University of Bari, Via E. Orabona, 4, 70125 Bari, Italy*

[c]*Artificial Brain S.r.l., Via Piave 63, 70125 Bari, Italy*

[d]*D.A.BI.MUS. S.r.l., Via Quintino Sella, 268, 70123 Bari, Italy*

ABSTRACT

Contemporary libraries have changed quickly their social role and function due to the proliferation and diversification of multimedia digital documents, becoming complex networks able to support communication and collaboration among the various distributed users communities. Technologies have not grown in step with the needs generated by this new approach, except in specific areas and implications. Hence the need to design an integrated digital library architecture that covers by advanced techniques the whole spectrum of functionality, without which the same social and cultural function of a modern digital library is at risk. This paper briefly describes an architecture that aims to bridge this gap, bringing together the experience, expertise and software systems developed by university and companies researchers. A prototype of the system is under development.

1. INTRODUCTION

Here Librarianship has evolved and taken new study and research fields in order to conceive and design new systems able to renew the management of information and, at the same time, to promote both cooperation between users and integration of heterogeneous information resources, ushering in the era of *Digital Libraries* (DL)[1]. Over the years, DL considerably has evolved from simple digital interface of public libraries' physical collections in complex networks, able to support communication and collaboration among different users communities worldwide. By contemporary DL citizens can access, discuss, evaluate and develop different types of information content[2]. Facing to the progress in research aimed at creation of DLMS able to support DL containing homogeneous digital collections and metadata, there are still hard difficulties in designing and implementing DL effectively integrating, managing and making accessible multimedia digital objects and related metadata. This paper proposes an "integrated" DL architecture, designed as a prototype for the project D_ISRAELI, an approved project under the "Living Labs Smart Puglia 2020" call by Apulia Region, presented by University of Bari Aldo Moro and CeRDEM

– Center for Research and Documentation in Judaism in the Mediterranean. The prototype, under development, aims at creating a DL on Jews history and culture in Apulia designed especially for librarianship and archival areas. The paper is structured as follow: section 2 presents a comparison about the major open source Digital Library Management Systems (DLMS) finalized to choose the best one to our aim. Section 3 details the components that should be integrated or enhanced in the choosen DLMS architecture in order to realize the set of innovative features identified by the component itself. Section 4 gives an overview of the major characteristics of the proposed DLMS, in particular the management of different types of metadata and text extraction and indexing from scanned digital objects. Section 5 gives the conclusions.

2. OPEN-SOURCE TECHNOLOGICAL BASE

The definition of the proposed architecture has required an analysis of available open source DLMS[3] in order to choose the one more suitable to our aims. Among the most known and used we have considered and compared *dSpace*, *EPrints* and *Greenstone*[4]. *Fedora Commons* has not been taken into account in this comparison because fundamentally it is a repository oriented to the digital data preservations rather than to the fruition. Furthermore, MARC protocol and the bibliographic data exchange protocol Z39.50 are not adequately supported by *Fedora Commons*, both non-negligible characteristics for the purposes of the DL to be realized. From the comparison between the three identified DLMS, the common features are: 1) OAI-PMH is supported, 2) storage and management of any type of content, 3) Multilingual Interface-oriented capabilities to the end user, and 4) production of statistical reports based on the count of records. Instead, they differ on the following characteristics:

- As unique identifier for the DL objects, dSpace uses CNRI Handle System (www.cnri.reston.va.us), Greenstone uses OAI Identifier (www.openarchives.org/OAI/2.0), while Eprints does not rely on any standard convention.

- In addition to the Dublin Core and METS metadata, common to all three DLMS, dSpace supports MARC/MODS too, while Greenstone NZGLS and AGLS.

- The search functionalities supported by dSpace (Field Specific, Boolean Logic and Sorting options) are a superset of those of EPrints and Greenstone.

- For the browsing functionality, EPrints and Greenstone allow the use of any field, while only Author, Title, Subject and Collection can be used with dSpace.

- User authentication in dSpace is possible via LDAP or Shibboleth, in Eprints only via LDAP, and User Groups in Greenstone.

- The databases that can be used are Oracle and PostgreSQL by dSpace, also MySQL and Cloud by EPrints, while Greenstone has an its own implementation.

- OAI-ORE, SWORD, SWAP are supported by dSpace (which adds SRW/U as extension of Z.39.50) and EPrints (which adds RDF), while Z39.50 by Greenstone.

The choice has fallen on dSpace essentially for the following three factors: 1) it supports better than the others the various metadata standards and protocols for interoperability, 2) it has been developed using only one programming language, and 3) it has a more complete documentation and there are various communities that provide support.

3. ARCHITECTURE OF THE PROPOSED SYSTEM

The proposed DLMS therefore extends the architecture of dSpace adding specific modules for the management of innovative features. The architecture is divided into three levels, according to the conceptual framework in Fig. 1. The Application Layer includes the following components for access to the system:

- Web UI is the module for the Web-based access both to the back-office area (via IDPs) and into front-end of the Digital Library through various portals. The back-office area allows the insertion and editing of digital content and its metadata, as well as managing users for access. The front-end allows visualization and rendering of contents through a Web interface that combines all advanced file formats guaranteeing multimedia, multichannel and protection. It also presents the front-end for the collaborative tagging (operated by Web 2.0 module).

- Mobile Devices allows viewing and rendering of contents through tablets and smartphones.

- Monitoring is the module that allows you to observe the behavior of the system and to produce reports in Excel format, XML, and PDF with the possibility of representation through graphs.

- The I/O module include interfaces that allow the metadata exchange using OAIS, OAI-PMH, Z39.50 and OAI- ORE. It also enable the rendering of content as Open Data.

- The Business Logic level includes the following modules suitable for the realization of the system functionalities:

- Core Tools is the module containing basic entities for system configuration and logging.

- Search Engine is the module that implements the functionality to support the information finding. Content indexing is done by the Lucene open source tool that implements techniques based on terms at the state of the art.

- Web 2.0 is the module for the management of collaborative tagging, i.e. it allows to perform the necessary functions to support the active participation of the portal users to the published contents.

- Access Management is the module for user authentication and profiling. It allows the access management both in normal mode or through IDP.

- Text Extractor is the module that enables the extraction of text from documents via ICRPad (see Sect. 4).

- Content Manager is the module designed to manage the objects, collections and licenses within the DL.

- Cataloging is the module that allows to manage digital contents that will populate the DL. The MAG/ICCU standard (version 2.0.1) compatible with standard METS will be used.

- Georeferentiation is the module that allows to reference the content in accordance with their geographic coordinates. This module will allow the storage and retrieval of content on the basis of spatial queries.

The Storage level manages access to the physical resources of the system deals with the organization of the content, including metadata, information about users and permissions associated with them, the status of the approval flow during the insertion of a content. Specifically: RDBMS Wrapper is the module that allows read/write access to the particular implementation of the DB (in this project is PostgreSQL, but is easily extendable to other types, e.g. Oracle). Bitstream Storage Manager is the module that permits the storage on the file system or SRB (backup tool). The DL operations will involve the various architectural modules. In addition to functionalities of a classical DL in the next section will be described some of the innovative aspects that will be integrated.

4. INNOVATIVE ASPECTS OF THE PROPOSED SOLUTION

To enable the efficient management and retrieval of content with different representation formats, we adopt a standard representation for administrative and descriptive metadata involving the use of different languages and hardly compatible with each other. To this end a good provision about the integration of tools that enable a separate management of metadata according to the various representation standards is necessary from the beginning. In order to facilitate interoperability, an extension of the dSpace basic architecture to support exchange protocols is required

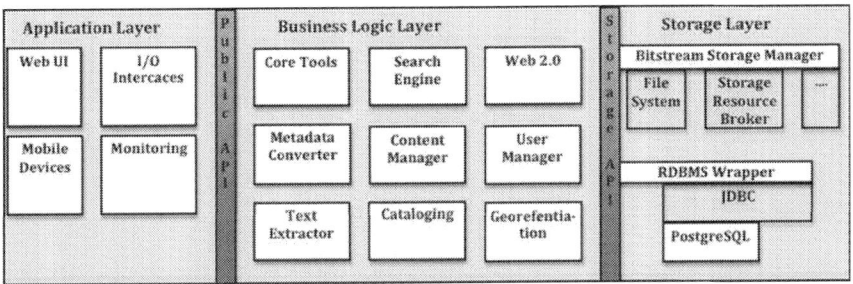

Figure 1. Architecture of the proposed system.

The standards supported by the DLMS interface require the integration of tools which ensure the proper functioning and approaches that permit the mapping among the different metadata representation usable in the standard interface. One of the most innovative aspects will be related to the high interactivity that the DL will ensure to different user communities in response to their cognitive demands. This will be made possible by the innovative feature on the indexing of extracted texts directly from the digital document images including manuscripts with, e.g., historical content. This feature will allow to build a search engine able to handle directly the information contained in these documents enabling the application of semantic indexing techniques. To this end, into some modules will be implemented features that are not natively supported by dSpace.

In *Web UI* module will be integrated recognition features, graphic matching and text extraction from digital objects into PDF/A with printed and manuscripts content using the digital platform for the recognition ICRPad[5]. It will support Text-based lexical indexing through state of the art technology provided by Lucene and those based on co-occurrences and semantics provided by the system DOMINUS[6], which also includes advanced features for the selective reading of interesting portions of digital format documents. The DLMS will support digital objects in different formats, in particular FITS (Flexible Image Transport System) for the images.

In *Profiling and Authentication* module will be implemented a user profiling behavior functionality aimed at customization of services: using advanced techniques developed in the Artificial Intelligence field for tracing the interactions of each individual user with services, it will make possible to infer specific information regarding various spheres such as the special interests, the preferences of interaction, the goals, routine activities, and more. Based on this information, customized service for each user will be provided, making its experience with the DL easier and more productive[7].

In *Cataloging* module will be included a common higher-level metadata schema based on Semantic Web languages. This will allow to take advantages offered by the abstract properties of the metadata during the search operation of the DL contents.

5. CONCLUSIONS

All Digital libraries have been radically changed by the increase of new technologies, evolving from mere digital counterpart of public libraries' physical collections in complex networks, able to support communication and collaboration among users communities worldwide. This increase created the need for integrated systems able to manage DL by advanced functionality, today still unsatisfied. This article has proposed an innovative DL architecture that aims to bridge this gap by new features, such as the integration of technologies for processing documents covering all phases: acquisition, content extraction, indexing, searching and enjoyment. A prototype system is being developed with very interesting prospects which could eventually lead to realize a DL model that functions as an integrated system for preservation, management and use of complex multimedia digital objects.

ACKNOWLEDGEMENTS

This work is partially funded by the project "VINCENTE – A Virtual collective INtelligenCe ENvironment to develop sustainable Technology Entrepreneurship ecosystems" (PON 02_00563_3470993) funded by the Italian Ministry of University and Research (MIUR).

REFERENCES

1. Arms WY. The1990s: The Formative Years of Digital Libraries. Library Hi Tech, 30(4), 2012, pp. 579 – 591.

2. Agosti M. Digital Libraries. *Mondo Digitale* settembre 2012;**43**:1-13.

3. Andro A, Asselin E, Maisonneuve M. Digital libraries: Comparision of 10 software. *Library Collections, Acquisitions, and Technical Services* 2012;**36**(3-4):79-83.

4. Tramboo SH, Shafi SM, Gul S. A Study on the Open Source Digital Library Software's: Special Reference to DSpace, EPrints and Greenstone. *International Journal of Computer Applications* (0975–8887) 2012;**59**(16).

5. Barbuti N, Caldarola T. An innovative character recognition for ancient book and archival materials: A segmentation and self-learning based approach. In Agosti M, Esposito F, Ferilli S, Ferro F, editors. Communications in Computer and Information Science. Vol. 354: *Digital Libraries and Archives*, IRCDL 2012, Heidelberg: Springer, (pp. 261-270).

6. Ferilli S, Esposito F, Basile TMA, Redavid D, Villani I. DOMINUSplus - DOcument Management INtelligent Universal

System (plus). In Agosti M, Esposito E, Meghini C, Orio N, editors. *Digital Libraries and Archives - Post-proceedings of the 7th Italian Research Conference* (IRCDL-2011), Communications in Computer and Information Science 249, 123-126, Springer, 2011.

7. Semeraro G, Costabile MF, Esposito F, Fanizzi N, Ferilli S. Machine Learning Techniques for Adaptive User Interfaces in a Corporate Digital Library Service. In *Machine Learning and Applications, Proceedings of the ACAI-99 Workshop W03 on Machine Learning in User Modeling*, 21-29, Chania, Crete, Greece, July 5-16, 1999.

CHAPTER 9

Damage function for historic paper. Part I: Fitness for use

Matija Strlič1*, Carlota M.Grossi1, Catherine Dillon1 , Nancy Bell 2 , Kalliopi Fousekil , Peter Brimblecombe3 , Eva Menart 1 , KostasNtanos2 , William Lindsay 2 , David Thickett 4 , Fenella France 5 and Gerrit De Bruin 6

¹Institute for Sustainable Heritage, University College London, London, UK.
²The National Archives, Kew, Richmond, Surrey, UK.
³University of East Anglia, Norwich, UK.
⁴English Heritage, London, UK.
⁵Library of Congress, Washington DC, USA.
⁶Nationaal Archief, The Hague, The Netherlands.

ABSTRACT

Background

In heritage science literature and in preventive conservation practice, damage functions are used to model material behaviour and specifically damage (unacceptable change), as a result of the presence of a stressor over time. For such functions to be of use in the context of collection management, it is important to define a range of parameters, such as who the stakeholders are (e.g. the public, curators, researchers), the mode of use (e.g. display, storage, manual handling), the long-term planning horizon (i.e. when in the future it is deemed acceptable for an item to become damaged or unfit for use), and what the threshold of damage is, i.e. extent of physical change assessed as damage.

Results

In this paper, we explore the threshold of fitness for use for archival and library paper documents used for display or reading in the context of access in reading rooms by the general public. Change is considered in the context of discolouration and mechanical deterioration such as tears and missing pieces: forms of physical deterioration that accumulate with time in libraries and archives. We also explore whether the threshold fitness for use is defined differently for objects perceived to be of different value, and for different modes of use. The data were collected in a series of fitness-for-use workshops carried out with readers/visitors in heritage institutions using principles of Design of Experiments.

Conclusions

The results show that when no particular value is pre-assigned to an archival or library document, missing pieces influenced readers/visitors' subjective judgements of fitness-for-use to a greater extent than did discolouration and tears (which had little or no influence). This finding was most apparent in the display context in comparison to the reading room context. The finding also best applied when readers/visitors were not given a value scenario (in comparison to when they were asked to think about the document having personal or historic value). It can be estimated that, in general, items become unfit when text is evidently missing. However, if the visitor/reader is prompted to think of a document in terms of its historic value, then change in a document has little impact on fitness for use.

KEYWORDS

Heritage management Conservation Heritage values Fitness for use Psychometrics Statistical experimental design (DOE)

1. BACKGROUND

In heritage management damage functions are used to model change in relation to environmental variables and other stressors, e.g. use. Damage functions are defined as 'functions of unacceptable change to heritage dependent on agents of change' [1], reflecting the fact that the term 'damage' reflects both the state of an object and the stakeholders' value of the level of change to that object that is unacceptable.

In our previous work [2], we quantitatively investigated the value users attach to library and archival heritage. The study involved 543 respondents in a variety of contexts: historic houses, reading rooms and exhibitions. Users were shown to have well-defined attitudes to objects, broadly characterised within nine categories: 'Future Value', 'Public Value and Evidence', 'Understanding the Present', 'Content and Learning', Personal Meaning and Identity', 'Discovery and Engagement', 'Rarity', 'Materials and Sensory Experience' and 'Connection to the Past'.

In the same work, we also investigated stakeholder attitudes to the future use of collections, and identified that ~90 % of users defined as visitors or readers, would want collection items to remain in a usable state (readable, or suitable for display) for the next 500 years. The most frequent response, around half of respondents, was 100 years. This corresponds to another study [3], where a similar question was posed to 187 experts curating or researching geological collections. This study reported ~70 % satisfaction with the same preservation horizon of 500 years. Other studies of perspectives on attitudes to the future have produced horizons of the same order of magnitude, if not necessarily the same particular horizon [4]. This is essential evidence for underpinning collection management protocols.

To effectively manage change of cultural heritage, we need to answer the question of how users interact with such change as well as whether different values affect such interaction.

This question is complex, as there are many processes of physical change, as well as values applied by users. It is also well known that such interactions might reflect the presumed 'use' of an object [5]. In this paper, we will define 'use' as the mode or context of use, for example display (objects are observed without handling) or reading, which involves manual interaction. There may be other purposes, depending on the type of interaction in question, including, e.g. storage.

The concept of use allows us to assess whether an object is fit for a type of use. As a term, fitness for use is similar to 'condition', although it does not include the susceptibility of an object to degrade as a criterion of condition. The concept of fitness for use is more meaningful because if an object is unfit for a particular use, a change in how it is accessed is required. For example it may need to be removed from an exhibition, or use may only be allowed under supervision.

This implies that benefits can no longer be accrued from interaction with such an object, meaning that its value for the particular type of use has been consumed and that the object reached the end of its lifetime (Fig. 1), unless an investment in a conservation intervention restores its fitness.

Since accrual of benefits occurs in the interaction between objects and users, it is useful to explore how users define the metrics of fitness. A number of methodologies have been developed for this purpose, most of which are based on psychometrics [6] and standard procedures exist for estimating aspects of image quality using psychometric scaling [7–9].

Psychophysics is the field of quantitative studies of perception, examining the relations between observed stimuli and responses. It allows for the relation to be extracted, while providing a reason for it [10] and has found application in different fields, from occupational and clinical health [11, 12], to politics and law [13].

Figure 1 Newsprint from 1918 to 1919, labelled as 'unfit for use' due to the fragile nature of the material

Psychophysical studies are broadly divided into two main classes. Threshold methods are based on the energy a user can just detect, such as the decibel level at which sound can be heard. These methods are used when detection of a stimulus is the factor of interest. Supra-threshold methods are used in instances where the stimulus is easy to perceive, and are used to discriminate between stimuli, such as discrimination between sounds [14].

The work on acceptable colour change focussed on threshold methods and the concept of 'just noticeable difference' (JND). JND represents "a stimulus difference that leads to a 75:25 proportion of responses in a paired comparison task" [15]. In assessment of image perception, this concept is used not only for colour changes, but all visually perceivable aspects of image quality e.g. blurring and contrast, as well as other aspects of image quality [16].

In the heritage field, this concept, including the synonymous term 'just perceptible colour difference', JND has mainly been used to understand the effect of illumination [17–20] despite the fact that the magnitude of a just noticeable difference remains ill-defined. If a more precise definition were in place [21], the timeframe for acceptable change could be defined as well. The concept of JND is well suited to measure the visual ability, rather than the viewer's perception of change in an image [22].

To determine fitness for use, concepts other than JND might be more useful, such as category scaling. The goal is to assign numbers to perceptual events, the benefit of which is that it allows for direct observer response to changes, although restricted to only a number of categories [23]. Category scaling methodologies exhibit high stability and lead to low participant stress [24]. Standard category titles have been developed which can be useful [25].

In this paper, we explore the attitudes of library and archival users to what could be defined as damage, i.e. the threshold at which change is no longer acceptable by users. We do this in the context of two types of use: reading and exhibition, and three contexts of value: pre-assigned personal or historic value, and no pre-assigned value.

2. METHODS

2.1 Value scenarios

To examine which values might affect the way users evaluate the fitness of an object, we used the outcomes of the VALUE questionnaire study

[2], which has clearly shown that different contexts of use are associated with:

- Different value profiles

- Different attitudes towards the future

- Different attitudes towards agents of change

It is evident that the value profile of a user may affect a fitness assessment and ideally, the influence of all types of values on fitness assessment might need to be evaluated. While from an academic point of view this might be justifiable, it is questionable whether the results could significantly affect collection management practice, particularly in large collections. It is therefore useful to examine only a few value types.

In this research we explored the potential to bias fitness assessment responses by describing the value of a document as a historical source and learning resource, without specifying the type of historic value (e.g. personal history, accountability, contextual, history), relating to the military, population migration, parish records, institutions including workplaces, hospitals and asylums, law and religion.

The second prompt was related to personal and community identity. To bias fitness assessment in this sense, the assessors were referred to lists of family member names, a relative's war record or information about a relative's early childhood, such as school or hospital records.

The third scenario was the 'random' scenario, where the assessors were given no specific information about the document they were assessing, thus providing un-biased data. In this scenario, the assessors considered the object as any random library or archival object, which allowed us to see how users assess fitness purely on the basis of material state of a document.

To summarize, we used the following three value scenario prompts:

1. *Random* "imagine you are assessing a random archival document"
2. *Personal* "Imagine that you are assessing documents you found during research into your family history"
3. *Historical* "Imagine that you are assessing documents by a historian interested in significant events in the history of your country.

2.2 Types of degradation

For psychometric scaling to allow for aspects of damage to be directly related to physical deterioration and to determine how this is affected by values, assessors were presented with a set of documents with deterioration features selected on the basis of the following criteria:

- Gradually accumulated during regular use (and not prior to acquisition or during catastrophic events, e.g. flood, fire, mould outbreak as such deterioration is the consequence of singular events rather than a continuous process)

- Easily visually detectable without prior expert material knowledge, thus allowing any archival user to participate.

In storage or display the mechanical strength or the colour of documents may change, however, it could be argued that only the latter can be easily assessed by a user without technical aids or expertise. On the other hand, during manual handling, tears and missing pieces might gradually accumulate as a consequence of reduced mechanical strength. Thus, the three aspects of degradation of interest are: tears, missing pieces, and discolouration.

Their influence on fitness for use could be assessed using actual archival documents, however, in such an experiment many variables could be difficult to control: document size, paper texture, writing style, deterioration prior to accession. Additionally, it would be difficult to prevent degradation of such archival documents during assessments.

Therefore, a suitable series of model documents was sourced from a notebook (1946) and therefore of identical appearance. These pages were further distressed to exaggerate the three aspects of deterioration (Fig. 2): discolouration, tears and missing pieces.

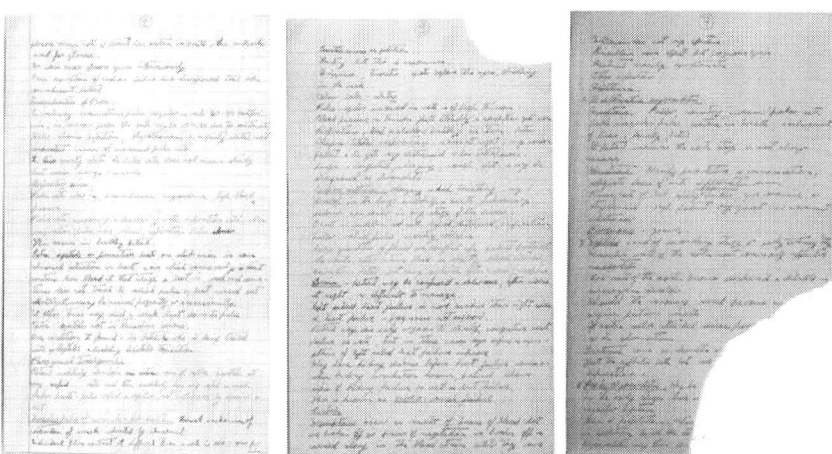

Figure 2 Examples of differently distressed documents used in fitness-for-use workshops, progressively discoloured and with a progressively big missing piece from *left* to *right*. The document on the *left* also has a large tear, stretching across text. The documents were written in English hence users tended to read them, which they were reminded not to in order for content not to affect their responses

Three levels of degradation were considered (Table 1). Discolouration was achieved by dry heat at 190 °C for 15 or 45 min.

2.3 Experimental design

The aspects of degradation could be studied separately: if a missing piece and a tear appear on the same page, the two aspects might be assessed independently or one relative to the other. For example, a highly discoloured page with a large tear and a small missing piece could be evaluated as more unfit than another document with a large missing piece only. Therefore, combinations of degradation aspects need to be examined, to see if the aspects add up or if they have synergistic (or antagonistic) interactions.

Table 1 Aspects and levels of degradation of the documents used for assessment of fitness

Level	Aspect of degradation		
	Tears	Missing piece	Discolouration ($\pm$SD, N = 5)
0	None	None	As acquired: $L^* = 97.3 \pm 0.2$, $a^* = -5.0 \pm 0.7$, $b^* = 12.6 \pm 1.5$
1	In the margin, not extending across the text	In a corner, not extending across the text	$\Delta E_{1976} = 21 \pm 6$, $L^* = 83 \pm 5$, $a^* = 0 \pm 2$, $b^* = 27 \pm 4$
2	Extending across the text ~1/3 into the page	Extending across the text, ~1/6 of the page missing	$\Delta E_{1976} = 36 \pm 7$, $L^* = 72 \pm 8$, $a^* = 5 \pm 2$, $b^* = 35 \pm 2$

Total colour difference is relative to level 0

To do so, the number of differently distressed documents used in a one-at-a-time experimental design would need to be 3^3, i.e. 27, which could make an assessment workshop long and potentially tiring for assessors.

To economise on the number of documents, while still being able to explore the interactions between aspects of degradation, principles of statistical Design of Experiments (DOE) were used. These allow for a significant reduction of the required experiments, as well as for variations of each aspect of degradation to be studied simultaneously, while the effect of each aspect on fitness for use can be evaluated independently. If the aspects have an additive effect, then DOE represents not only a more economic experimental design, but also allows for more precision; however, if the aspects are not additive, but interact synergistically, then DOE can allow for detection and estimation of such interactions [26]. The advantage of DOE is that it allows for the maximum amount of information to be extracted using the minimum number of experiments, and for efficient handling of experimental errors [27, 28].

The simplest are factorial experiments, where all factors are varied simultaneously at a limited number of factor levels [26]. More complex DOEs involve response surface designs, such as the central composite design (CCD), which is among the most popular types of response surface designs. It consists of a factorial design with centre points, augmented with a group of axial points that allow for estimation of curvature. Such a design requires only 15 differently distressed documents, while the centre point can be repeated several times for a better estimation of the uncertainty. Three such repetitions give the total number of documents of 17. The DOE response surface designs and data analysis were carried out using Minitab 16 Statistical Software (State College, PA, USA).

The three factors of degradation, i.e. tears (T), missing pieces (MP) and discolouration (D) now need to be assigned to the three factors to produce the documents with required combinations of aspects of degradation (Table 2) at three levels: 0, 1 and 2.

The documents, single loose handwritten pages, were distressed in such a way that no aspect of degradation was removed or obscured by any other. As documents assessed for the purpose of reading were to be handled to reflect the normal process of reading, they gradually degraded during the workshops, in which case they were replaced.

2.4 Fitness-for-use workshops

In accordance with "Methods" it was examined whether thresholds determined for a particular type of use can be manipulated by the relationship that people form with the document under the influence of value factors. Hence, prior to the workshops, users were prompted to consider that the documents they were observing have a personal or historical significance, or not.

Table 2 Documents required for the fitness-for-use assessment workshops, with combinations of aspects of degradation: tears (T), missing pieces (MP) and discolouration (D), as required by the CCD response surface design, with three repetitions of the centre point

Document no.	T	MP	D
1	0	0	0
2	2	0	0
3	0	2	0
4	2	2	0
5	0	0	2
6	2	0	2
7	0	2	2
8	2	2	2
9	0	1	1
10	2	1	1
11	1	0	1
12	1	2	1
13	1	1	0
14	1	1	2
15	1	1	1
16	1	1	1
17	1	1	1

To account for the two types of use (display and handling) and three value scenarios (*Random*, *Personal* and *Historical*), six variations of workshops were developed (Table 3).

Participants were volunteers drawn from among the public at the workshop locations. All participants were given equal information about the purpose of the workshops, before any assessment was carried out. Further information was provided only after they completed the assessments. To control this, workshop assistants were trained and given the same information on how to interact with assessors.

Table 3 Six fitness-for-use workshop scenarios, accounting for two different types of use and three value scenarios, with the associated codes, as used in the analysis

Code	Value scenario	Use
H–D	*Historical*	Display
H–R	*Historical*	Reading with handling
P–D	*Personal*	Display
P–R	*Personal*	Reading with handling
R–D	None (*Random*)	Display
R–R	None (*Random*)	Reading with handling

Per scenario, 50 participants were required. The workshops were conducted at The National Archives (Kew, UK), Library of Congress (Washington DC, USA), and Wellcome Library (London, UK).

The objects were viewed individually, without taking other documents as a reference and participants did not move objects to compare one to another. The objects were not displayed in the numeric order (1–17, as in Table 2), but were arranged randomly.

For the two aspects of use, the participants were encouraged to interact with the documents in the following way:

1. Reading under the normal conditions as are expected in libraries and archives, where a document can be freely handled and read

2. Viewing, under the normal conditions as are expected in exhibition spaces, where a document is presented in a way that it is intelligible without handling, separated from the viewer.

The medical manuscripts usefully allowed for association with a historic event or a personal circumstance, however, since the fact whether or not a user could read the handwriting might affect their assessment, they were discouraged to read.

Each participant was asked to categorize the 17 documents for their fitness for the specified purpose. Categories from 1 to 5 were modified on the basis of those prescribed by BS ISO 20462-2:2005:

1. 'Excellent'

2. 'Good'

3. 'Quite good' (in the UK context) or 'Reasonably good' (in the US context)

4. .'Not good'

5. 'Unfit'

The data collected across the six groups of participants and scenarios were expected to indicate the threshold of damage, as represented in the various combinations of deterioration features in each of the sets of documents. Personal and historic value scenarios (H–R, H–D, P–R, P–D) were thought to provide an indication of how thresholds of damage are influenced by the two specific value contexts. In the following section we analyse the strengths and limitations of this approach.

RESULTS AND DISCUSSION

The premise of the workshops was that users have determinable acceptance levels for states of documents when considered for particular use and with specific attached values, and that these acceptance levels can be revealed through the fitness-for-use workshops. The acceptance levels represent thresholds of damage, whereby users combine their visual observation of a physical state with a judgement of what state is required for a particular type of use. Conversely, damage is defined as loss of fitness for use (unacceptable change).

Across the three participating institutions, The National Archives, Library of Congress and the Wellcome Library, 331 users responded and carried out the assessments during November and December of 2012 (Table 4). This provided a dataset slightly larger than required by DOE.

Table 4 Total number of collected response sheets per scenario and institution

Workshop scenario	The National Archives	Library of Congress	Wellcome Library	Total
H–D	23	40	0	63
H–R	24	5	19	48
P–D	25	41	0	66
P–R	24	5	15	44
R–D	24	42	0	66
R–R	24	5	15	44

For workshop scenario abbreviations see Table 3

At the Library of Congress, the workshops were mostly carried out in the exhibit room, which provided an ideal context for the display scenarios, and only 15 responses were collected in the reading rooms. On the other hand, in the Wellcome Library, all of the responses were collected in the reading rooms, providing the context for scenarios H–R, P–R and R–R. At The National Archives, the workshops were carried out in a space which was separated from the reading room and from the exhibition space, so both modes of use could be explored.

Since the participants were asked to provide a written description of what they imagined the documents to represent (cf. "Value scenarios"), it was of interest to analyse whether the value prompts were effective. As a preliminary test, the responses were arranged into word clouds (Fig. 3), where the frequency of a word is represented by the size of the font used to depict the word in the cloud.

Figure 3 Word clouds for responses to the question "What did you imagine the document to represent?" for the workshops scenarios H–R and H–D (*above*) and P–R and P–D (*below*)

It appears that most assessors looked for a connection between the textual content of the document and a historic event or a personal circumstance. In the case of H–R and H–D, the most frequent associations were with the Civil War (in the US context) or with WWI or WWII (in the UK context). In the case of P–R and P–D, the key words are family, diary, grandfather, letters, history and war, indicating that the prompts were potentially successful. This indicates that participants responded positively to the value suggestions given to them by workshop assistants, who helped if the participants expressed difficulties with suitable associations. An analysis of the *Historical* and *Personal* value scenarios could thus be meaningful.

Frequency analysis of the expressed fitness for use (FfU) responses was performed on the individual response data, i.e. using all of the individual responses corresponding to the individual documents, per workshop scenario (Fig. 4). It is of interest to note that most responses are skewed towards the lower numbers, meaning that in general the users thought that the documents were on average 'Good' to 'Quite good' ('Reasonably good'). This is perhaps most evident in the case of H–D and R–D scenarios, and less so in the R–R scenario, and could indicate that when users were not given a value scenario they are the least forgiving.

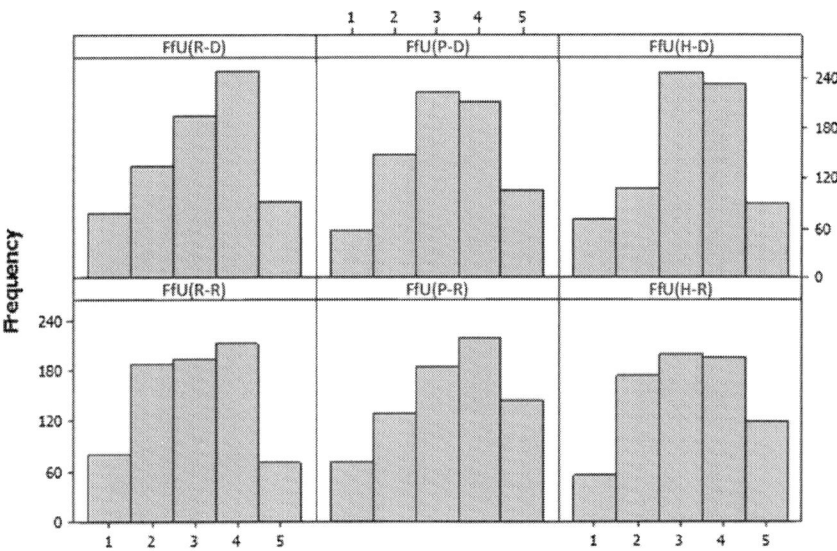

Figure 4 Frequency analysis of responses (*FfU* fitness for use), where each document represents one response (i.e. 17 per participant), per scenario. The ratings were *1* 'Excellent', *2* 'Good', *3* 'Quite good'/'Reasonably good', *4* 'Not good', *5* 'Unfit'

A linear comparison of responses (FfU) vs. individual aspects of degradation (T—tears, MP—missing pieces, D—discolouration) was performed (Fig. 5) to investigate if there is one single aspect that has most influence on fitness for use. It is evident that it is likely that MP has the most pronounced effect on fitness as the regression between MP and fitness is strongest for all six scenarios, while the regressions between T and D, and FfU are weak.

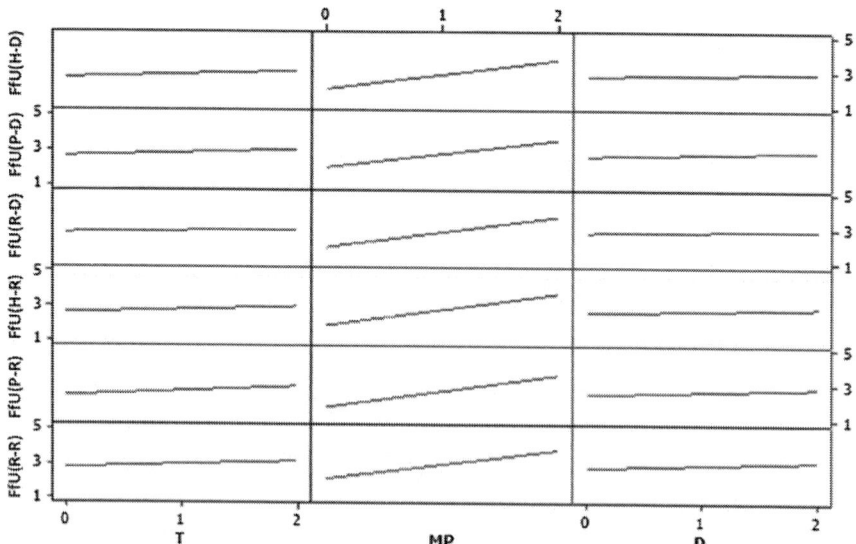

Figure 5 Regression lines between FfU values (1–5) and levels of degradation (1–3) for tears (T, *left column*), missing pieces (MP, *middle column*) and discolouration (D, *right column*), for the six scenarios: H–D, P–D, R–D, H–R, P–R, R–R, from *top* to *bottom*

Further quantitative analysis was performed to obtain the response surface in the form of the following general surface equation (Eq. 1):

$$FfU = c_1 T + c_2 MP + c_3 D + c_4 T^2 + c_5 MP^2 + c_6 D^2 \\ + c_7 T \cdot MP + c_8 T \cdot D + c_9 MP \cdot D + Const.$$ (1)

In this equation, terms 1–6 represent linear and quadratic terms representing the three aspects of deterioration without any interaction, while terms 7–9 represent interaction terms. If the interactions are significant, then coefficients c_7, c_8 and c_9 should be statistically significant ($p < 0.05$).

After the response surface model is developed on the basis of actual participant responses (FfU), the actual responses can be compared with the calculated (modelled responses), and the quality of the correlation between actual and modelled FfU values can be explored using linear regression, and expressed as R^2, i.e. the squared regression coefficient— the closer it is to 1, the better the model.

The developed response surfaces have R^2 values (Table 5) that might be considered low for physical sciences, but indicate a moderate-large effect size according to Cohen's conventions, often cited in the psychology literature [29]. It is possible to appreciate that certain models better describe the responses, as the R^2 values are higher.

Table 5 R^2 values for the different surface response models, for the 6 workshop scenarios

Scenario	FfU model	
	Equation 1	MP only
H–D	0.3817	0.3562
H–R	0.4533	0.4029
P–D	0.3907	0.3693
P–R	0.4536	0.3836
R–D	0.4191	0.4017
R–R	0.4148	0.3569

The coefficients modelled on the basis of Eq. 1 are compared for the different workshop scenarios in Fig. 6. It is evident that there are a few consistent features across all the scenarios. There is no statistically significant interaction between discolouration (D) and missing pieces (MP), except in one scenario, R–R. Discolouration seems to be evaluated mostly independently of the other aspects of physical degradation, and significantly contributes to only one scenario, i.e. R–R, reading of documents of no pre-assigned value.

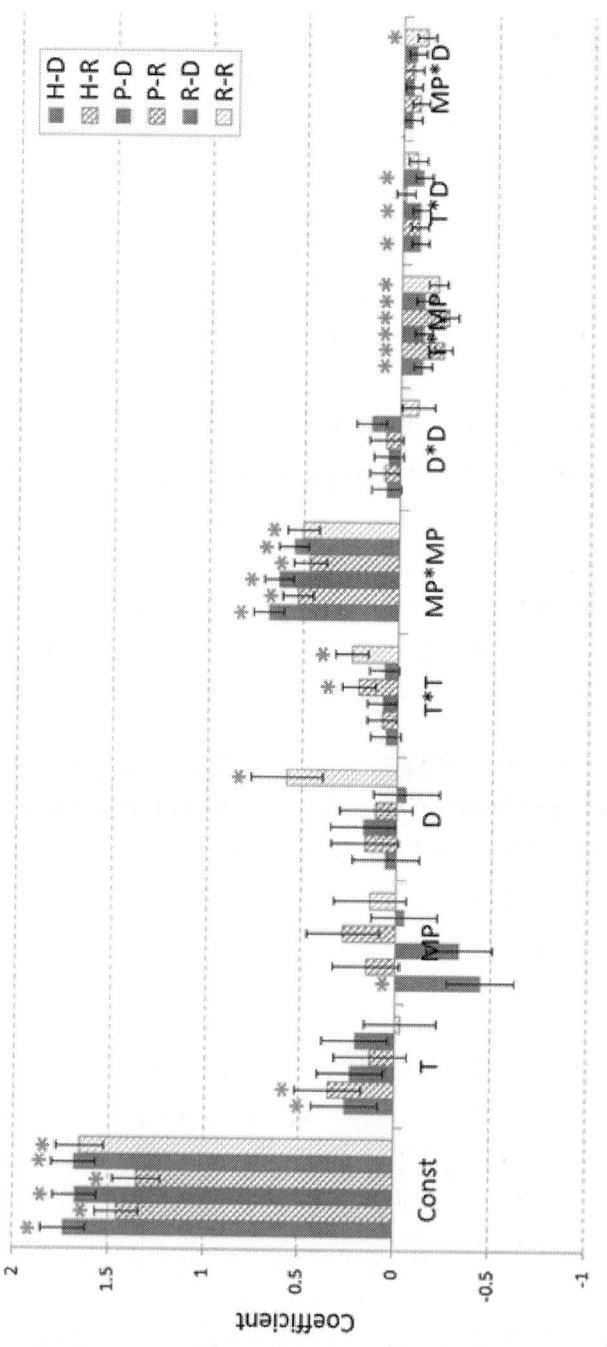

Figure 6 Values of FfU response surface coefficients based on Eq. 1, with the associated uncertainties, for the six workshop scenarios. *Stars* indicate terms of statistical significance ($p < 0.05$)

The other interactions, $T \times D$ and $T \times MP$ are mostly statistically significant, although they contribute little to the overall score, as the coefficients are small. It seems therefore that tears and missing pieces seem to be also mainly observed as independent aspects of degradation.

Another interesting observation is that tears mostly contribute insignificantly to the overall score, although they seem to be more important in the context of manual handling (scenarios H–R, P–R, R–R). Readers seem to be concerned with tears if they need to handle an object: tears have an impact on how a reader holds a document—e.g. how easy it is to turn a page or pick a document up without damaging it further.

Overall, it is evident that the quadratic term $MP \times MP$ contributes most to the overall score, indicating that users are mostly concerned with the textual content of a document. This is in agreement with the data presented in Fig. 5.

We therefore modelled the responses by using only the terms MP and $MP \times MP$. The quality of this model is less good than the one based on Eq. 1, though only marginally so in most scenarios, as evident from the regression coefficients in Table 5. In Fig. 7, we can examine the differences between the coefficient values for the six scenarios. All are statistically significant.

The *Historic* value scenario significantly differs from other scenarios in that both the Constant and the MP term coefficient are significantly different from the other four scenarios, while the quadratic term coefficient is similar for all of them. With the linear term being negative for the H–D and the H–R scenarios, this means that the FfU responses for these value scenarios will be overall more similar, leading to the conclusion that with a historic document in mind, users mind least about how distressed the documents are, regardless of the access context, i.e. display or reading.

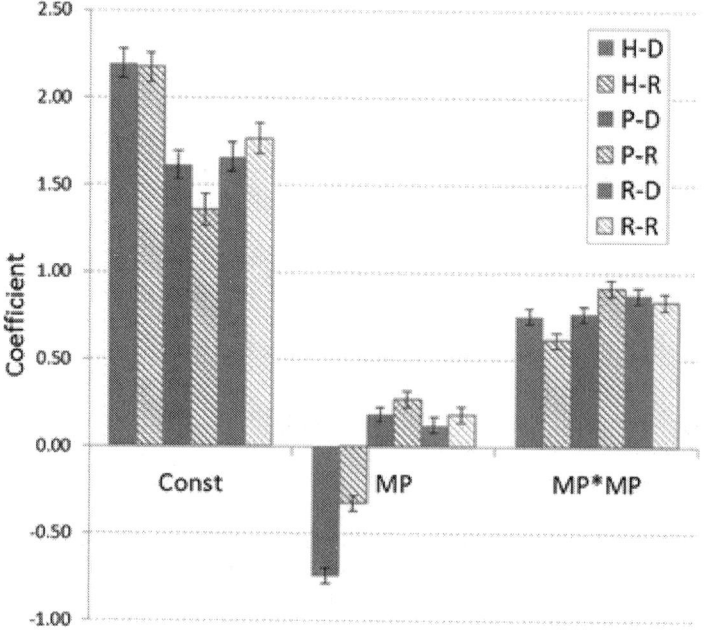

Figure 7 Values of FfU response surface coefficients (quadratic model, MP only), with the associated uncertainties, for the six workshop scenarios. All terms are statistically significant ($p < 0.05$)

There seems to be a further minor, although statistically significant difference between H–D and H–R, i.e. for display purposes, the aspects of degradation are least important and most documents were considered 'Good' or 'Quite good' ('Reasonably good'), as is also evident from the corresponding frequency plot in Fig. 4.

The other scenarios, P–D, P–R, R–D and R–R, are similar. This leads to the conclusion that users were unable to identify with the personal value prompt, and this is no longer taken into account in the analysis to follow.

The conclusion is that although tears contribute to fitness to a small extent, the model based on MP as the most important aspect of physical degradation to general library and archival users, as presented in Fig. 7, explains most variance in the data. This allows us to easily estimate the level of degradation at which users assess a document as not being fit for use (Fig. 8).

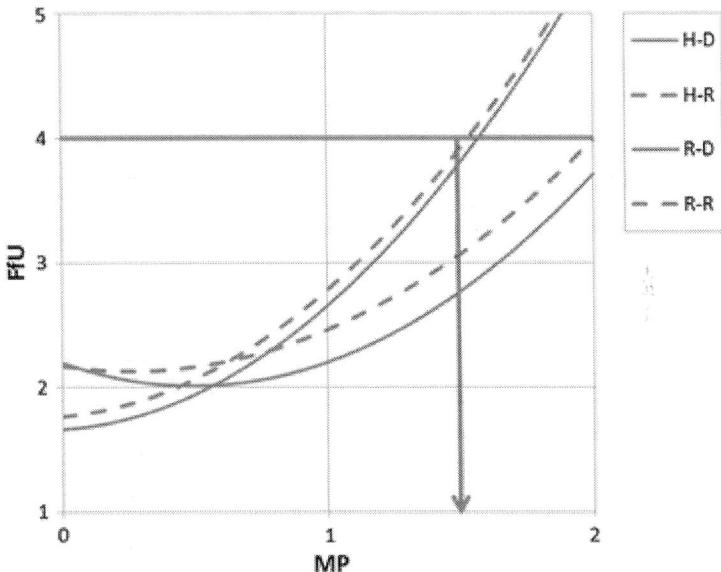

Figure 8 Models representing fitness for use (FfU) depending on the extent of the page missing (*MP* missing piece). The *red lines* enable a graphical assessment of the critical MP at which users would start assessing the document to be unfit

However, scenarios H–D and H–R show a different picture to the rest. Regardless of the size of the missing piece (the largest missing piece was ~1/6 of a page), assessors never consider a historic document to be unfit, regardless of the context of use, i.e. display or reading. In the context of collection management, the physical state of a document of historic significance is not considered to be particularly important to its fitness.

On the other hand, if the document was considered to be a random archival document, the point at which users assess it as unfit for use can be defined, and is estimated graphically in Fig. 8. This point is the same, regardless of the context of use, i.e. display or reading.

With a missing corner of a page not containing any text, most users were of the opinion that the document was 'Quite good'. However, if the text was evidently affected at which the state of a document became unacceptable (i.e. damage [1]). At this point, the value of MP is ~1.5 (Fig. 8). While on the basis of this experiment it is not possible to estimate how much missing text this level of deterioration represents, the estimation that any missing text will significantly affect the level of satisfaction is meaningful.

CONCLUSIONS

We explored attitudes of users to visually observable material change in paper documents, with the aim of understanding what extent of degradation is characteristic of objects that are no longer fit for use in the specific contexts of display or reading with handling, and when different types of values are elicited. In collection management practice, such extent of degradation could be seen as the end of lifetime for the particular mode of general access, and environmental and access practices could be adapted to optimise the lifetime as required. By taking the views of users into account we also make collection management more publically accountable.

Fitness for use was assessed in user workshops, where 331 participants were confronted with a number of differently distressed objects. The aspects of degradation explored were those of interest to practical collection management: gradually accumulating and possibly preventable or, at least, slowed down. This ensures that the assessed fitness thresholds are of practical significance and applicability. Fitness is affected by a number of aspects, such as:

- The values, reflected in the attitudes of assessors to the objects.

  If an archival object is perceived to be of historical value, even large missing pieces do not make it unfit for some uses, as opposed to those documents for which no value is elicited. It is questionable, however, whether the value modalities can be taken into account in the management of large collections of objects of similar significance due to resource issues.

- The aspects of degradation studied.

  For archival documents, changes in colour and tears contribute to the overall assessment to a minor extent, whereas missing pieces contribute most.

- The purpose, or context of object use.

  There are statistically significant differences between how fitness is assessed for archival documents that are intended to be displayed and for those intended to be read. In the latter case, aspects of mechanical degradation (tears, missing pieces) are more important.

- The document's information content.

  In the case of documents for which no particular value is elicited, obviously missing text in a document leads to that document being assessed as unfit for use.

- Having defined the threshold fitness for use for paper-based documents, and the long-term planning horizon in which this might be acceptable, a dose–response function is required that enables us to calculate the accumulation of tears and missing pieces as a consequence of frequency of use, as well as loss of strength as a consequence of natural ageing. We will explore this in Part II of this series of papers.

ABBREVIATIONS

CCD: central composite design; DOE: design of experiments; FfU: fitness for use

TYPES OF PHYSICAL DETERIORATION

T: tears; MP: missing pieces; D: discoloration

WORKSHOP TYPES

- H–R: workshop scenario assuming the object being of Historical (H) value, assuming the object being read while manually handled (R— Reading);

- H–D: workshop scenario assuming the object being of Historical (H) value, assuming the object being read from a distance (D—Display);

- P–R: workshop scenario assuming the object being of Personal (P) value, assuming the object being read while manually handled (R— Reading);

- P–D: workshop scenario assuming the object being of Personal (P) value, assuming the object being read from a distance (D—Display);

- R–R: workshop scenario assuming the object being of no pre-assigned, Random (R) value, assuming the object being read while manually handled (R—Reading);

- R–D: workshop scenario assuming the object being of no pre-assigned, Random (R) value, assuming the object being read from a distance (D—Display).

AUTHORS' CONTRIBUTIONS

MS, CMG, CD, NB, KF, PB, EM, KN, WL, DT jointly developed the concept of this work. CMG, CD, KN, DT, FF, GDB organised the workshops and developed guidance for assessors. MS, CMG, CD and EM produced the samples. MS, CMG and CD analysed the data. All co-authors contributed to the manuscript. All authors read and approved the final manuscript.

ACKNOWLEDGEMENTS

The authors are grateful to colleagues who helped organising workshops at The National Archives (Kew, UK): Carolien Coon, Hannah Clare, Dinah Eastop, Richard Williams; Library of Congress (Washington DC, USA): Cindy Connelly Ryan, Meghan Wilson, Matthew Kullman, Ray Privott, Mark Sweeney, Dianne van der Reyden, Hans Wang; Wellcome Trust Library (London, UK): Gillian Boal, Carolien Coon, Stefania Signorello, as well as at the Nationaal Archief (The Hague, The Netherlands), where a preliminary test workshop was organised: Gabrielle Beentjes and Roberto Padoan.

We are particularly indebted to 331 volunteers who took part in the fitness for use workshops.

COMPETING INTERESTS

The authors declare that they have no competing interests.

REFERENCES

1. Strlič M, Thickett D, Taylor J, Cassar M. Damage functions in heritage science. Stud Conserv. 2013;58:80–7.

2. Dillon C, Lindsay W, Taylor J, Fouseki K, Bell N, Strlič M. Collections demography: stakeholders' views on the lifetime of collections. In: Ashley-Smith J, Burmester A, Eibl M, editors. Climate for collections conference, Munich, Doerner Institut, 7–9 November 2012, Postprints. London: Archetype; 2013. p. 45–58.

3. Robb JE. Quantitative assessment of stakeholder attitudes to geological collections for improved collections management, MRes Thesis, UCL Centre for Sustainable Heritage. 2012.

4. Lindsay W. Time perspectives: what 'the future' means to museum professionals in collections-care. The Conservator. 2005;29:51–61.

5. Jones S, Holden J. It is a material world: caring for the public realm. London: DEMOS; 2008.

6. Clark K. From significance to sustainability. In: Clark K, editor. Capturing the public value of heritage—the proceedings of the London conference, 25–26 January 2006. Swindon: English Heritage; 2006. p. 59–60.

7. Mason R. Assessing values in conservation planning: methodological issues and choices. In: De la Torre M, editor. Assessing the values of cultural heritage. Los Angeles: J. Paul Getty Trust; 2002. p. 5–30.

8. Cossons N. Capturing the values of places—opening remarks. In: Clark K, editor. Capturing the public value of heritage—the proceedings of the london conference, 25–26 January 2006. Swindon: English Heritage; 2006. p. 70.

9. Throsby D. The value of cultural heritage: what can economics tell us? In: Clark K, editor. Capturing the public value of heritage—the proceedings of the London conference, 25–26 January 2006. Swindon: English Heritage; 2006. p. 40–3.

10. Jowell T. From consultation to conversation: the challenge of better places to live. In: Clark K, editor. Capturing the public value of

heritage—the proceedings of the London conference, 25–26 January 2006. Swindon: English Heritage; 2006. p. 7–13.

11. Langford M. Colour and conventions. In: European colour photography. London: The Photographer's Gallery Ltd. 1978.

12. Engeldrum PG. A theory of image quality: the image quality circle. J Imaging Sci Technol. 2003;48:447–57.

13. Baird JC, Noma E. Fundamentals of scaling and psychophysics. Wiley Series in Behavior. New York: Wiley; 1978.

14. Ehrenstein WH, Ehrenstein A. Pyschophysical methods. In: Windhorst U, Johansson H, editors. Modern techniques in neuroscience research. Berlin: Springer; 1999.

15. British Standards Institution. BS ISO 20462-1:2005(E) Photography—psychophysical experimental methods for estimating image quality. 2006.

16. Bradley JS, Reich R, Norcross SG. A just noticeable difference in c-50 for speech. Appl Acoustics. 1999;58:99–108.

17. Ashley-Smith J, Derbyshire A, Pretzel B. The continuing development of a practical lighting policy for works of art on paper and other object types at the Victoria and Albert museum. In: Bridgland J, editor. ICOM-CC 13th Triennial meeting, Rio de Janeiro. London: James & James Ltd.; 2002. p. 3–8.

18. Crawford BH. Just perceptible colour difference in relation to level of illumination. Stud Conserv. 1973;18:159–66.

19. Michalski S. The lighting decision. In: Fabric of an exhibition, preprints of textile symposium. Ottawa: Canadian Conservation Institute. 1997. pp. 97–104.

20. Brokerhof AW, Reuss M, MacKinnon F, Ligterink F, Neevel H, Fekrsanati F, Scott G. Optimum access at minimum risk: the

dilemma of displaying Japanese woodblock prints. In: Conservation and Access: 2008 IIC Congress, London. 2008.

21. Derbyshire A, Ashley-Smith J. A proposed practical lighting policy for works of art on paper at the V&A. In: Bridgland J, editor. ICOM Committee for Conservation 12th Triennial Meeting, volume 1, Lyon. London: James & James Ltd.; 1999. p. 38–41.

22. Richardson C, Saunders D. Acceptable light damage: a preliminary investigation. Stud Conserv. 2007;52:177–87.

23. Ehrenstein WH, Ehrenstein A. Pyschophysical methods. In: Windhorst U, Johansson H, editors. Modern techniques in neuroscience research. Berlin: Springer; 1999.

24. British Standards Institution: BS ISO 20462-1:2005(E) Photography—psychophysical experimental methods for estimating image quality. 2006.

25. British Standards Institution: BS ISO 20462-2:2005(E) Photography—psychophysical experimental methods for estimating image quality. 2006.

26. Box GEP, Hunter JS, Hunter WG. Statistics for experimenters : design, innovation, and discovery. 2nd ed. Hoboken NJ: Wiley; 2005.

27. Yamini Y, Saleh A, Khajeh M. Orthogonal array design for the optimization of supercritical carbon dioxide extraction of platinum(IV) and rhenium(VII) from a solid matrix using cyanex 301. Separ Purif Technol. 2008;61:109–14.

28. Vanaja K, Shobha Rani RH. Design of experiments: concept and applications of Plackett Burman design. Clin Res Regul Aff. 2007;24:1–23.

29. Cohen J. A power primer. Psychol Bull. 1992;112:155–9.

CHAPTER 10

Damage function for historic paper.
Part II: Wear and tear

MatijaStrlič1*, Carlota M.Grossi 1 , Catherine Dillon 1 , Nancy Bell 2 ,
Kalliopi Fouseki 1 , Peter Brimblecombe 3 , Eva Menart 1 , Kostas
Ntanos 2 , William Lindsay 2, David Thickett 4, FenellaFrance 5 and
GerritDe Bruin 6

1 *Institute for Sustainable Heritage, University College London, London, UK.*
2 *The National Archives, Kew, Richmond, Surrey, UK.*
3 *University of East Anglia, Norwich, UK.*
4 *English Heritage, London, UK.*
5 *Library of Congress, Washington, DC, USA.*
6 *Nationaal Archief, The Hague, The Netherlands.*

ABSTRACT

Background

As a result of use of library and archival documents, defined as reading with handling in the context of general access, mechanical degradation (wear and tear) accumulates. In contrast to chemical degradation of paper, the accumulation of wear and tear is less well studied. Previous work explored the threshold of mechanical degradation at which a paper document is no longer considered to be fit for the purpose of use by a reader, while in this paper we explore the rate of accumulation of such damage in the context of object handling.

Results

The degree of polymerisation (DP) of historic paper of European origin from mid-19th–mid-20th Century was shown to affect the rate of accumulation of wear and tear. While at DP > 800, this accumulation no longer depends on the number of handlings (the process is random), a wear-out function could be developed for documents with DP between 300 and 800. For objects with DP < 300, one large missing piece (i.e. such that contains text) developed on average with each instance of handling, which is why we propose this DP value as a threshold value for safe handling.

Conclusions

The developed model of accumulation of large missing pieces per number of handlings of a document depending on DP, enables us to calculate the time required for an object to become unfit for use by readers in the context of general access. In the context of the average frequency of document use at The UK National Archives (Kew), this period is 60 years for the category of papers with DP 300, and 450 years for papers with DP 500. At higher DP values, this period of time increases beyond the long-term planning horizon of 500 years, leading to the conclusion that for such papers, accumulation of wear and tear is not a significant collection management concern.

KEYWORDS

Heritage management Conservation Fitness for use Wear and tear Mechanical properties Reliability engineering

1. BACKGROUND

In continuous incremental degradation processes, such as degradation by chemical reactions whose rate is influenced by heat and humidity, or due to the effect of light/UV or pollution, material change is most often modelled deterministically. Examples of such modelling are dose–response functions, and the most recent review for historic paper was published in 2011 [1]. These presume that by knowing previous states of the material and all model parameters, material properties can be uniquely determined at any point in time. All objects that are exactly the same thus degrade in precisely the same way under the same conditions. This is a reasonable assumption that can be used in modelling, provided that we know the most important physical properties of historic objects that affect their degradation.

Physical degradation, i.e. degradation leading to mechanical failure as a consequence of excessive or repetitive mechanical stress e.g. due to handling, is rarely studied in collections. A notable exception is environmentally-induced mechanical stress, particularly in relation to objects made of wood and painted objects, where humidity fluctuations lead to internal stresses that result in cracking and other forms of mechanical failure [2, 3].

The probability of mechanical failure increases as a result of constant physical and chemical degradation of materials as they become weaker and therefore more susceptible to damage by fatigue and fracture. Incremental degradation of materials does not result in a sudden change, but it does increase the probability of failure due to fatigue. The reliability engineering concept described as the so-called 'bathtub curve' [4, 5] conceptualises these phenomena showing the rate of failure in a population of objects as a function of time and consisting of three phases. Of interest here will be the second and the third phase, when wear and tear accumulates either at a constant rate, or possibly at increased rates as the flexibility of cellulosic fibres decreases with loss of their degree of polymerisation (DP). It is useful to stress that mechanical properties of certain paper types, such as coated or heavily sized, may significantly depend on additives, however, such paper types are not of interest to this research.

For paper-based material, brittleness is often discussed as the most important consequence of environmentally induced chemical degradation. A number of collection surveys performed since 1970s onwards, have relied on a fairly straightforward, although destructive, manual folding test [6]. In this test, a page corner is folded as many times it takes for it to break off on gentle pulling.

The 1988 Swedish study [7] was one of the most thorough early studies, and used sixfolds as the criterion to categorise paper into the most degraded, with "many tears and/or pieces missing, worn edges, slight to strong yellowing, breaking corner on sixfolds". Using these criteria, it was estimated that the proportion of books with paper in the worst state was approximately the same in the National Archives of Sweden and in the Uppsala University Library, i.e. 20 %. These were mainly from the period 1860–1890, which coincides with the introduction of mechanised rosin sizing across much of Europe in 1850s [8]. In similar studies at the Library of Congress, 30 % of collections were estimated to be no longer fit for the purpose of manual handling, and similarly, 40 % in the Harvard University Library and 26 % in the Stanford University Library [9]. Other recent studies [10] continue to report similar values.

The premise of such studies is not based on an understanding of what users might perceive as an unfit document [11], but rather on the assessment of the current 'conservation state' of the document and, in some cases, of the risk of damage by handling, as perceived by conservation experts. Such perceived risk, however, still needs to be quantitatively justified by a study showing how wear and tear accumulates during use. Additionally, some of the above mentioned surveys are based on assessment elements, which users clearly do not perceive as unacceptable change, e.g. yellowing or discolouration, tears or minor missing pieces [11].

There have been many attempts to correlate mechanical properties of paper with a more easily measurable quantity. It has been estimated that with a degree of polymerisation of 250, paper becomes too brittle even to be tested for its tensile strength [12]. In a recent study in collaboration between the National Archives of the Netherlands, Sweden and Slovakia and the Royal Library of the Netherlands [13], an attempt was made to quantify the folding test results and sixfolds were found to correspond to a DP of approximately ~350. The uncertainty associated with the manual test was found to be significant, so interpretation of the result requires some caution.

Further related values have recently been reported. In a study of iron gall ink drawings, conservators assessed that the risk of drawings becoming damaged during handling by non-expert users was too high at DP ~ 400 [14]. In a study of cellulose-based painting canvases, involving a panel of 19 conservation experts, it was estimated that the risk of painting transportation and handling became too high if the DP < 400 [15].

However, neither the minimum number of folds nor the corresponding DP value, as determined by conservation experts, reflect how wear and tear is accumulated during handling of books and archival objects, i.e. the 'material wear-out' phase of the bathtub curve [4].

Such a wear-out function for library and archival materials would need to be able to predict the point of loss of fitness for use for general access, i.e. in a reading room or an exhibition (e.g. reading with handling, or display), discussed in the first paper in this series [11]. It was shown that neither discolouration nor tears matter to users, while deterioration considered unacceptable for the purpose of reading (with handling) or display is related to missing pieces of text. It is therefore necessary to study how missing pieces accumulate during the use of paper with different mechanical properties.

Previous research has demonstrated a correlation between DP and mechanical properties such as tensile strength and fold endurance [16]. While highly brittle paper documents continue to retain value as 'museum objects' even after their life as an object of general utility has been spent [17], it is important to stress that our interest is in library and archival use in general access conditions, i.e. without supervision.

Paper of very low DP has low fold endurance and tensile strength and is experienced by readers as brittle and difficult to handle. Therefore, it could be assumed that there is an association between DP and the occurrence of physical changes associated with mechanical properties over a given number of challenges (e.g. tears, loose sheets, missing pieces and folds which result in breaks). It could also be assumed that there is an association between DP and the extent of use required before such changes become classed as damage. Such a function, combined with the frequency of use in an experimental case study, may reveal that regardless of its brittleness, a document could continue to be fit for a considerable length of time if the frequency of its use (reading, display) is low. In other words, if a document is not used at all, its brittleness is not a concern. This paper thus addresses the accumulation of wear and tear.

2. METHODS

2.1 Experimental design

For the purpose of the case study, a set of mock documents (bound, i.e. books, and unbound, i.e. loose sheets) made from historic paper of known chemical and mechanical properties were subjected to physical challenges representative of handling that might be typical of a historic archive or library collection. After rounds of physical challenges (ten instances of handling each), the documents were assessed for changes to their physical state, i.e. deterioration. Of particular interest were tears and missing pieces, as their influence on user evaluation of fitness for use was assessed in our previous research [11].

The experimental design was expected to yield the following types of data:

1. The number or degree of changes per challenge, i.e. ten instances of handling (e.g. X number of large or small tears, Y number of large or small missing pieces).

2. Given that the challenges were intended to be representative of handling in a real library or archival setting, and the frequency of use in archives/libraries is generally known, then the above data can be converted to expected number of changes per period of time.

3. The above data and analysis was obtained for bound and unbound documents and was used to explore the comparison between a typical library document and a typical archival folder.

The SurveNIR Reference Historic Paper Collection [18] was used as the source of paper. The textbooks, guidebooks and fiction books range from mid-nineteenth to mid-twentieth century, and are written in Slovenian, Croatian, Serbian, German and Italian and 25 were selected for the simulation study on the basis of the DP of the paper in the books (~300– 1000) and the amount of paper available, as both a bound and an unbound object had to be produced using the same paper. The paper was characterised in terms of variables such as molecular weight, DP, pH, fibre furnish and other properties, using traditional analytical techniques, such as viscometry for DP and cold extraction for pH [19]. Since some of the papers used in this study contained lignin and could not be dissolved in the solvent used for viscometry, their DP was determined using size exclusion chromatography by Morana RTD d.o.o. (Ivancna Gorica), following the previously described methodology [20]. The average uncertainty of DP determination was ~10 %, which included experimental and sampling uncertainties and results of this research need to be interpreted in light of this uncertainty.

It also needs to be clarified that all papers used in this study were made following the European papermaking practices of the time (rosin sizing), acidic and not coated. All were printed.

From among the total of 25 pairs of bound/unbound simulated objects, 4 were additionally degraded in HCl-enriched atmosphere at 80 °C overnight, in order to rapidly obtain highly brittle material. The DP of these papers was measured using the same technique as outlined in [19], following ISO 5351:2010. As this research is only concerned with the material state (measured in terms of DP) at the time of the experiment described in "Challenges", previous environmental history (or indeed ultra-accelerated degradation in HCl-atmosphere) of the experimental papers is not of interest.

From each SurveNIR book, a 0.5–0.7 cm stack of paper (several sections) was rebound by a professional bookbinder (sections hand-sewn, spine glued with PVAc, hard binding on board, including an end paper, mull spine lining, fabric cover over spine, uncovered, with overhang, paper trimmed around the edges). This type of binding enabled all books to be comparable and data independent of binding type.

Most of the books had 6 sections, giving 48 sheets or 96 pages per document. For one book there was only enough paper to rebind 2 sections, giving 16 sheets or 32 pages. For two bound documents there was only enough paper to rebind 3 sections, giving 24 sheets or 48 pages.

For each of the 25 bound documents, there was a matching document made of unbound paper from the same SurveNIR sample book, with sections cut along the inner margin to create sets of loose sheets, which were additionally trimmed around the borders. Using guidance on storage of loose-sheet documents provided by The UK National Archives, Kew (TNA), the unbound documents have been placed in four flap folders. The unbound documents and folders have been labelled with their original SurveNIR code and the pages have been renumbered. Figure 1 shows some examples.

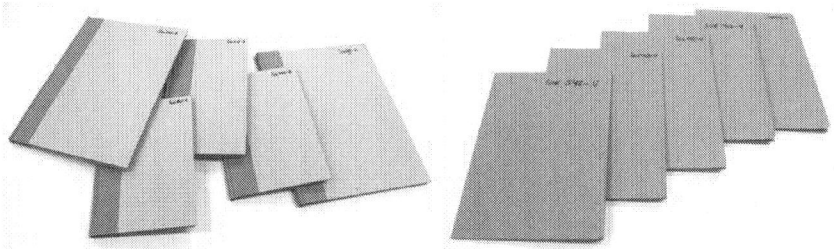

Figure 1 Examples of bound sample books and unbound paper in folders, as used in the study

Eight sets of three pairs (bound and unbound, same paper) of documents were compiled, with one pair of documents set aside as replacement if needed. In each set of three pairs there was a range of paper DP values (low, medium and high DP pairs) in order to avoid confounds of DP with any physical challenges.

Each set of documents was packaged in a large envelope that prevented wear and tear during transportation.

3. CHALLENGES

The documents were subjected to blocks of physical challenges. Each challenge had the following components:

- Packing.

- Tasks performed by participants, involving transport to the challenge venue (office or home), unpacking, performing a task with each document, repacking and transport back to the lab.

- Assessment of each document by the researcher–assessor.

For a block of physical tasks (i.e. a challenge) the aim was to subject each document to 10 instances of handling. A number of challenge cycles was conducted, mostly 9, representing 90 instances of handling. Each challenge on a given document was performed by a different participant. Their brief was to leaf through the document 10 times as if reading/searching for information. Some paper was too brittle and allowed for only 5 challenges before assessments of change became very onerous due the extent of physical degradation (Fig. 2).

Figure 2 The consequences of five challenges (50 instances of handling representing reading with handling) for the most brittle paper

To simulate reading with handling, participants were asked to leaf through an entire document in a short space of time. It took about 1–2 h to complete the tasks for a set of documents. The 9 participants were members of staff, researchers and volunteers (non-experts) at UCL Institute for Sustainable Heritage.

To make the study as representative as possible, the participants were assigned to challenges using a random number generator. Additionally, in each challenge, a participant received three books and three archival folders with paper of low, medium and high DP. No participant was assigned the same object twice.

Wear and tear could also occur during transportation of documents within an institution. However, the characteristics of transport and handling occurring in real historic libraries and archives vary from institution to institution, and potentially within an institution. For example, an archival folder might be retrieved from the repository, used very briefly by a reader, and then returned to the repository on the same day. Another box might be retrieved and used intensively for a week before being returned to the repository. In addition, different parts of a collection are held in different locations and produced to different reading rooms. Because of this, potential mechanical deterioration between the challenges was avoided by packing the documents into sturdy envelopes.

4. ASSESSMENT

Assessment took place at the end of every block of physical challenges. The key measured features followed the elements of distress as discussed in Part I [11]: tears and missing pieces. These were classified according to size and location:

- Small: tears of less than 5 mm from the edge of a sheet.

- Medium: tears or missing pieces in the 5–20-mm boundary, or tears that affect the inner margin in a book (binding edge).

- Large: tears or missing pieces that cross over the 20-mm boundary, within which text/images would be affected.

It is important to note that the assessment method used in this study is not 'condition' assessment [5] but rather a quantitative measure of physical change, specifically designed to enable categorisation in terms of the fitness of a document.

Since tears and missing pieces are caused by handling, their occurrence should be correlated with the mechanical properties of paper. Therefore the assessment measures are intended to be both indicators of use and sensitive to paper mechanical strength.

To avoid assessor bias, the assessment methods were developed and piloted in collaboration with several assessors and researchers initially. A number of books were independently assessed by two assessors on two occasions. To standardise the assessments, the following tools were developed:

- Templates. These were necessary as the sheets were trimmed to remove any previous wear and tear, which was assumed to have an effect on the experiment. Two assessment templates were designed: a template for bound documents and a template for unbound documents, printed onto a transparency and placed over the sheet being assessed.

- The template divides a sheet into several regions: A–H (Bound) and A–I (Unbound). These are overlaid on a 1-cm grid used for measuring the length of tears and the area of missing pieces. The red-dotted line marks the 5-mm boundary around the edge of a sheet (within which features are only counted). The solid red line represents a 2-cm margin (Fig. 3).

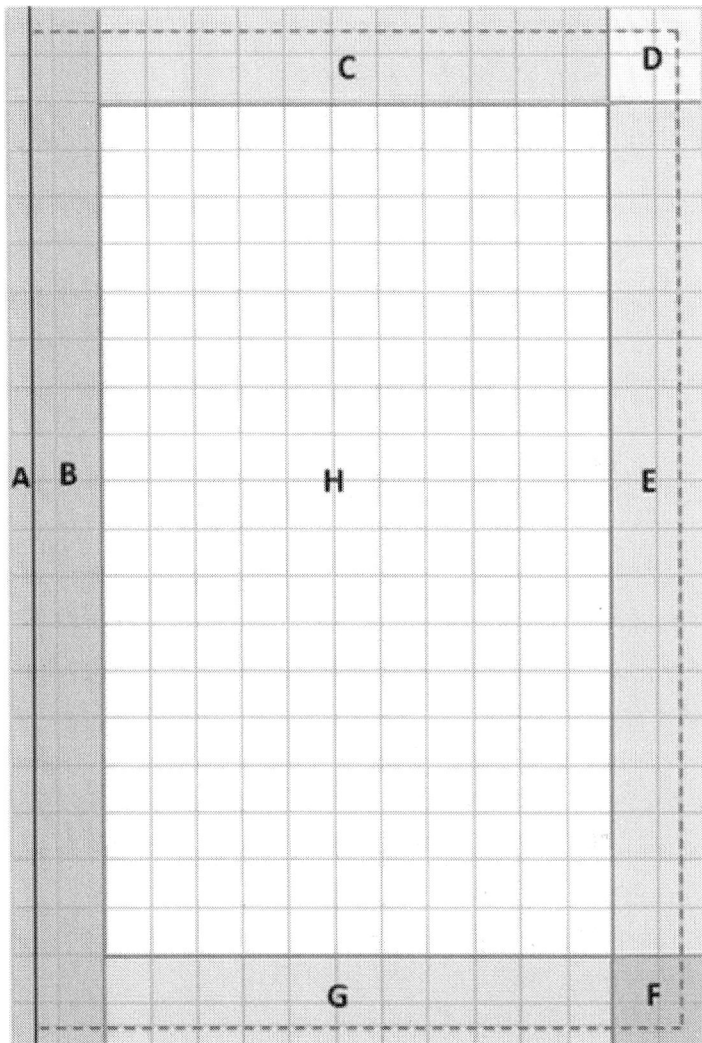

Figure 3 The template used in the assessment of wear and tear for bound volumes *boundary A* represents the inner (*binding*) margin

- The template represents a typical sheet from a book based on measurements taken before rebinding (e.g. most bottom margins were >2 cm wide), but could be moved around in order to assess sheets of differing dimensions.

- A template excel file was created to record the data and each spreadsheet had the following elements:

- Sheet: features were recorded sheet by sheet.

- Tears: tears around the edge of the sheet within the 5-mm boundary were counted (likely to be approximate); other tears were recorded individually (location, length, direction etc.). Since small tears can develop into large tears and large tears can lead to missing pieces and, therefore, the elimination of tears, changes in tear numbers could be positive as well as negative.

- Missing pieces: similarly to tears, small missing pieces were counted and larger ones were described in detail, categorised and summed up.

Based on piloting, there was greater agreement between assessors about the location of a feature and whether the text area was affected, rather than finer details of features (e.g. whether a tear has affected the text and images in a document, rather than how long and in what direction the tear is). These are also likely to be the features most important to readers when assessing fitness for use [11]. Hence, these were the key measures used to describe change in the documents over time.

Following the pilot assessment phase, all changes (tears, missing pieces) in all document sets were assessed by the same researcher after each challenge.

5. FREQUENCY OF DOCUMENT USE AT THE NATIONAL ARCHIVES, KEW

Data on the frequency of use were made available by The National Archives, Kew, per bay (set of shelves) and not down to any greater detail, e.g. shelf, box or reference number. The data illustrate that yearly productions of individual documents would be expected to be very low, with 50 % of the c. 20,000 bays at Kew having documents produced from them between 1 and 10 times per year (Fig. 4).

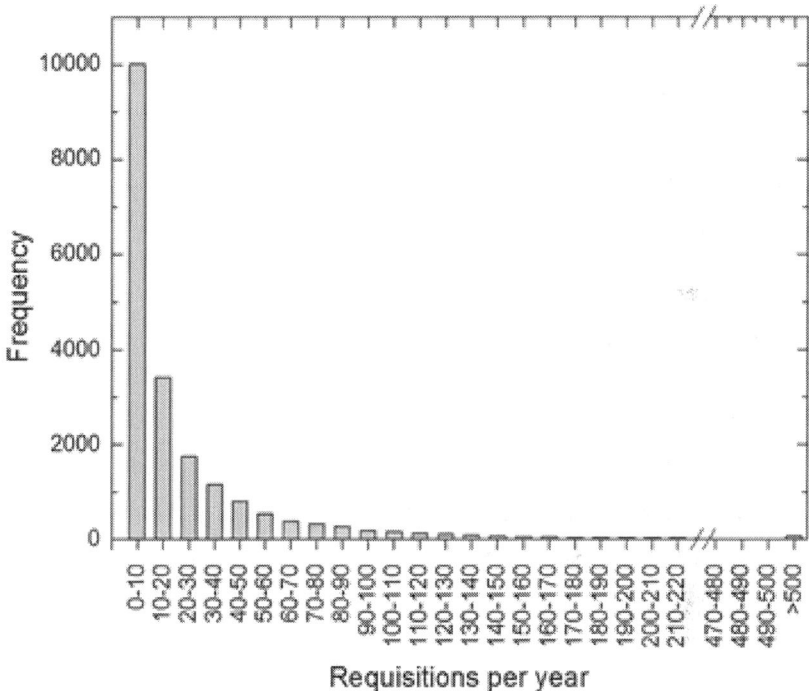

Requisitions per year

Figure 4 Frequency *plot* representing the number of uses (requisitions) per bay, per year, at The National Archives, Kew. Documents from 10,019 bays are requested 0–10 times per year, and from 77 bays >500 times per year

6. RESULTS AND DISCUSSION

6.1 Overall deterioration frequency analysis

Nine participants were involved in the experiments, carrying out a challenge per week. The introduced wear and tear was comparable across the nine participants, as demonstrated in Fig. 5. The plot represents average fractions (percentages) of deterioration in an item they 'treated' compared to all observed deterioration in the same item, caused by all of the individual participants.

It is of interest to note that in some cases the introduced deterioration was negative, which is reasonable since in the calculation of total deterioration, medium and large tears and missing pieces were added up. Since a medium tear may transform into a large tear or a large tear into a missing piece, different aspects of wear and tear may accumulate or transform into other aspects of deterioration and thus disappear. Small tears were not considered in the calculation as in our previous study [11] it was shown that it is predominantly large elements of degradation that lead to loss of fitness.

Since not all participants treated all the documents, and some may have treated the more fragile ones and others not (the most fragile four documents were only treated 50 times, all others 90 times), bias could have been introduced in the data. However, the data in Fig. 5 shows that it is difficult to systematically discriminate between the participants. While P1 introduced most deterioration on average, it is quite clear that the scatter of data is high. While P6 on average introduced less deterioration, second only to the most careful P7, the maximum deterioration they caused was the second most intensive.

6.2 Accumulation of mechanical deterioration

Interestingly, wear and tear accumulates approximately linearly with the number of challenges, as shown in Figs. 6 and 7. This again indicates that there is probably insignificant difference between the participants and the force they applied to objects. More importantly, this allows us to describe the data and develop functions showing the accumulation of aspects of deterioration as a function of the number of challenges (and by interpolation, the number of times an object was handled, resembling requisitions in an archival or a library context). The linear relationship between aspects of deterioration and the number of challenges also indicates that accumulation of wear and tear can be treated as an incremental process.

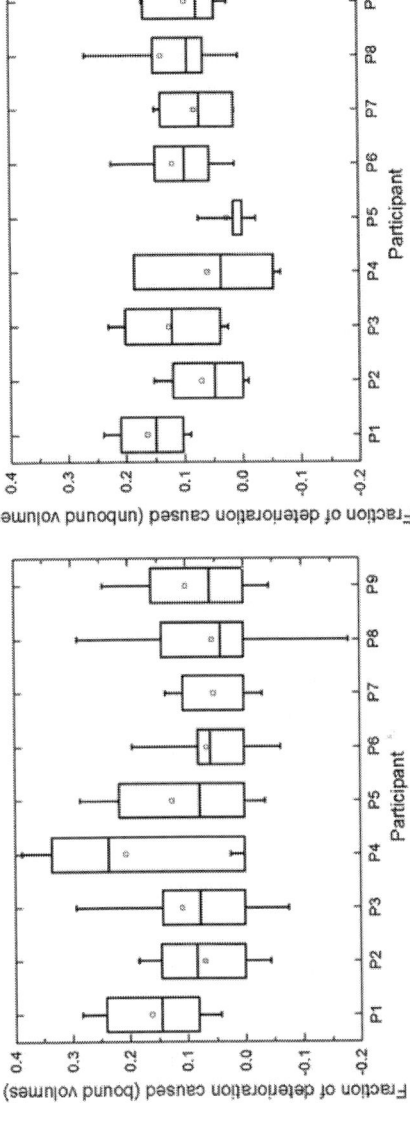

Figure 5 *Boxplots* comparing the fractions of the total number of elements of deterioration (small and large, tears and missing pieces) per document, caused by individual participants (P1–P9). *Left* bound (books); *right* unbound documents (archival folders). A *boxplot* shows the means (*open square*), the standard deviation (*whiskers*), whereas the *boxes* represent the first and the third quartile and the median. *Left* bound documents; *right* unbound documents

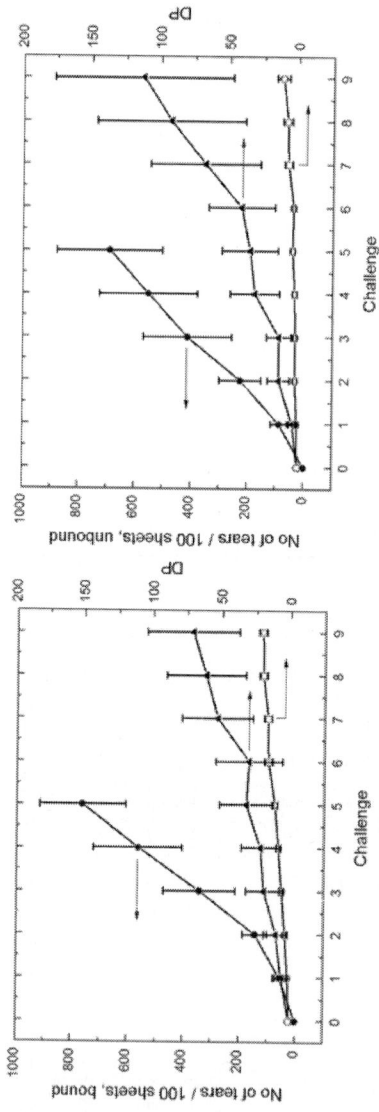

Figure 6 Accumulation of tears (all sizes) calculated per 100 sheets during the 5–9 challenges (50–90 treatments/instances of handling). *Left* bound volumes, *right* unbound folders. The *left y-axis* shows values for the DP interval: *triangle* – DP < 300 (N = 5), the *right y-axis* for all other DP intervals: *circle* – 300 < DP < 600 (N = 5), *open square* – 600 < DP < 900 (N = 7). For DP > 900, no mechanical deterioration was caused. The *error bars* represent the standard deviation for the average number of tears observed in the individual DP interval

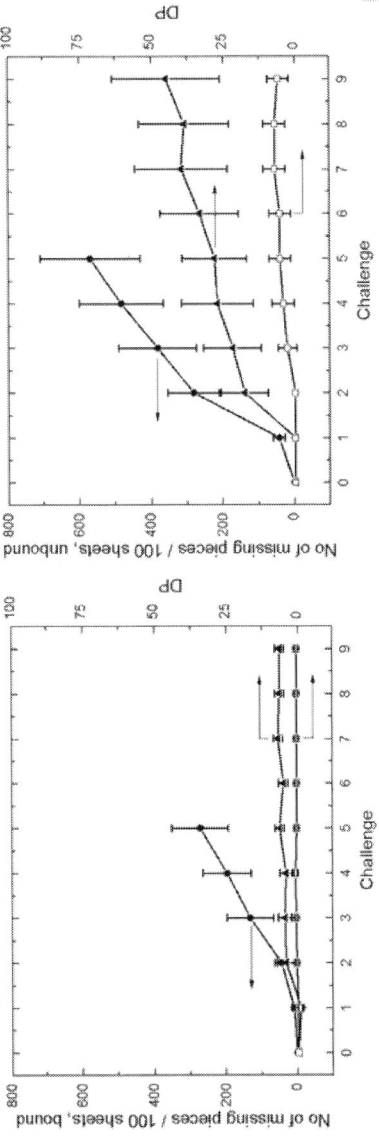

Figure 7 Accumulation of missing pieces (all sizes) calculated per 100 sheets during the up to 9 challenges (90 treatments). *Left* bound volumes, *right* unbound folders. The *left y-axis* shows values for the DP interval: *triangle* – DP < 300, the *right y-axis* for all other DP intervals: *circle* – 300 < DP < 600, *open square* – 600 < DP < 900. For DP > 900, no mechanical deterioration was caused. The *error bars* represent the standard deviation for the average number of missing pieces observed in the individual DP interval

It is of interest to note that tears develop at a similar rate in bound and unbound items, while missing pieces develop at a much lower rate for bound items than unbound ones. Although this observation indicates that for books and archival items different dose–response functions could be developed, all the observations were used in the following calculations (Fig. 8) due to data scatter. It is also possible that further differences could be found for object of different size, weight or type of binding; however, this was outside the scope of this research.

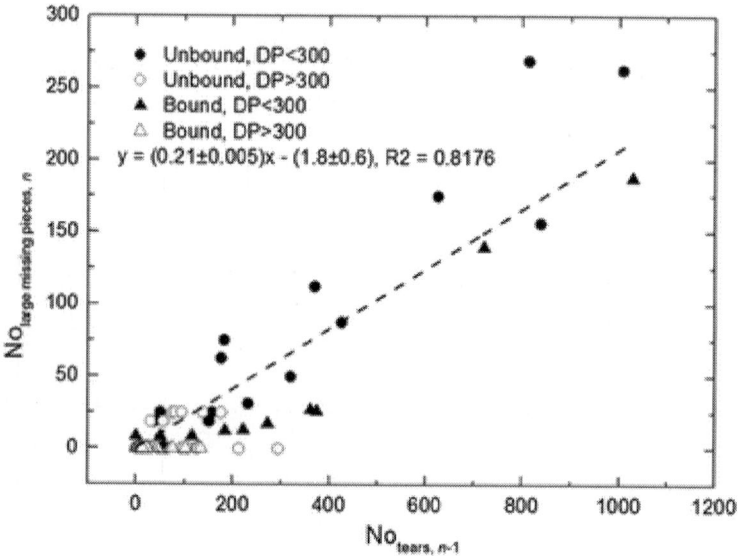

Figure 8 Accumulation of large missing pieces in the challenge number n, as a function of the number of all tears in the preceding challenge (n-1), both per 100 sheets, for bound and unbound volumes. The *regression line* is drawn using all data points

The finding that missing pieces accumulate more slowly than tears is also useful, and it might be reasonable to assume that missing pieces develop as a function of tears. This turns out to be the case and in Fig. 8, the number of large missing pieces accumulated during a challenge is plotted as a function of the number of tears in the preceding challenge, but only with two DP categories: DP < 300 and DP > 300, as the number of missing pieces accumulating in paper with DP > 300 is very small in comparison.

This is an interesting finding, as according to the threshold fitness for use, discussed in our previous paper [11] it is a large missing piece which makes a document unfit (unless a user has attached a strong historic value to the document, in which case such deterioration is generally viewed with more tolerance).

Extending this definition to a collection of sheets in a library or archival item (with potentially several hundred sheets) is possible by assuming that all textual information in the item is of equal importance.

In practical terms and on average, a large missing piece does not occur in isolation. In a document of 100 sheets, the first large piece will go missing after approximately 5 tears accumulate in a given sheet, which is already quite extensive wear and tear, not unlike the examples depicted in Fig. 2.

Appreciable accumulation of missing pieces (defined here as one per handling or more, per 100 sheets, where 'handling' refers to one 'instance of handling' as used in this study, i.e., leafing through the entire document once as if reading/searching for information) only occurs in objects with DP < 300: only 2–5 % of all missing pieces accumulate in paper with DP > 300, all the rest (95–98 %) in paper with DP < 300 (Fig. 7). The value of DP 300 could therefore serve as a useful threshold value of DP for the estimation of risk of accumulation of missing pieces and reaching the end of fitness for reading/handling, which is in agreement with the literature discussed in the introduction.

6.3 Wear-out function

Based on the collected data, we can now develop functions looking at both how large missing pieces develop as a function of the DP of paper as well as with the number of handlings or time. The latter depends on the frequency of production in an archive and for this, we need data on the frequency of production, as discussed in "Frequency of document use at the National Archives, Kew". The wear-out function represents the final period of the bathtub curve [4, 5] leading to loss of fitness.

In a real environment it may take many years before objects begin to lose large missing pieces, if the frequency of production is low, and especially if DP > 300. Therefore, to assess the period of time that might be needed to lose a large missing piece, both the frequency of production and the DP of paper will play a role.

Even objects with very low DP will not deteriorate mechanically (i.e. develop tears and missing pieces) if they are not handled—the period of time until loss of fitness for use in the context of general access in a reading room will therefore be extremely long (if not 'indefinite'). However, it is worth stressing that the process of wear and tear accumulation and the process of chemical deterioration (DP loss) are here treated as independent, which works in most cases: if objects are already very deteriorated (DP < 300) then further DP loss will not increase the rate of wear and tear accumulation, and if the objects are not very deteriorated (DP > 500) then the rate of accumulation is insignificantly low in comparison. It is only for objects that are highly accessed and nearing the threshold of DP 300 that the two functions would need to be used in conjunction.

In Fig. 9 (left), we show data on the rate of documents becoming unfit, i.e. developing at least one large missing piece per 100 sheets, as they are handled (leafing through the entire document once as if reading/searching for information). The cases where objects did not develop a missing piece, were not taken into account in this calculation (i.e. all objects with DP > 800). The data for bound and unbound objects are combined in these calculations, and a logarithmic function has been used as a model due to the large observed differences.

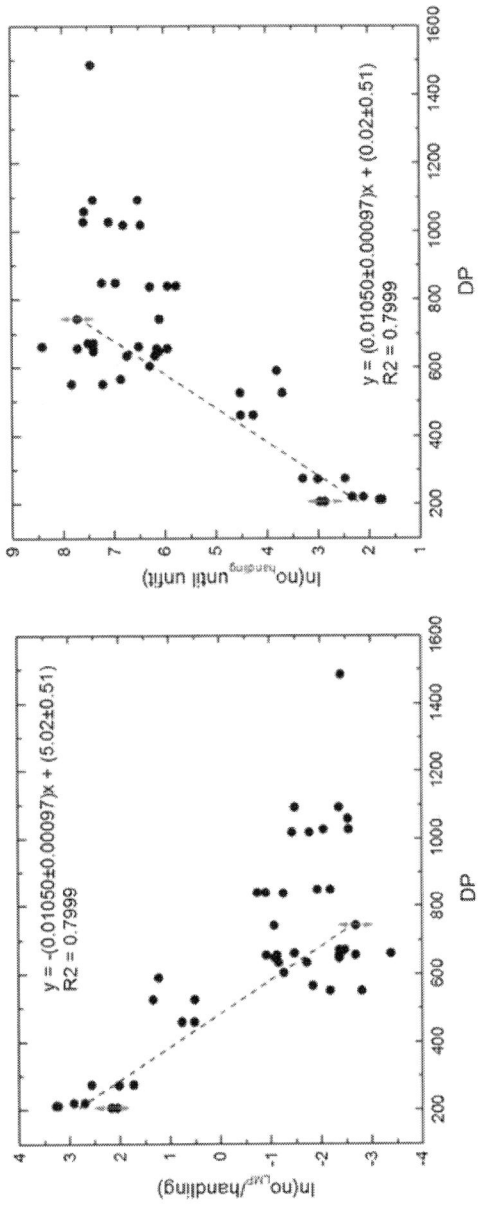

Figure 9 *Left* accumulation of large missing pieces (text missing) per handling, i.e. leafing through the entire document once as if reading/searching for information, of paper (per 100 sheets) depending on the DP of paper. *Right* number of handlings until such paper develops one large missing piece per 100 sheets, depending on their DP. *LMP* large missing pieces (i.e. a piece with text missing)

Figure 9 (right) shows the number of handlings until a document becomes unfit for use, depending on the DP of paper. In order to characterize the state of 'unfit for use' a threshold number of missing pieces per 100 sheets needs to be defined. This is a collection management decision, and in this calculation, one large missing piece per 100 sheets was taken as the threshold of fitness.

The equations describing functions in Fig. 9 are:

$$\ln\left(no_{LMP}/handling\right) = -(0.01050 \pm 0.00097)$$
$$\cdot \, DP + (5.02 \pm 0.51)$$

and

$$\ln\left(no_{handling}\,until\,unfit\right) = (0.01050 \pm 0.00097)$$
$$\cdot \, DP + \ln(threshold_{LMP}/100\text{sheets}) - (0.02 \pm 0.51)$$

where DP is 200–800. For $DP > 800$, the value no longer increases, indicating that the process of accumulation of wear and tear no longer depends on DP. In reliability engineering terms, this would relate to the 'random failure phase'.

With this last equation, it is now possible to calculate the average time required for a document with a given DP to deteriorate to the point where it accumulates one large missing piece per 100 sheets (Table 1).

It is evident that for documents with $DP > 500$, wear and tear is generally not an important issue, as the period of time until loss of fitness is comparable or longer to the planning horizon [11, 21]. However, for objects with $DP < 500$, wear and tear increases with importance as the DP decreases. It is important to note that even low-DP objects still survive a lot of handling, and that given how infrequently most documents are handled, there is very little loss of fitness even for documents that are considered fragile (DP < 300).

Table 1 Period of time until documents with a given DP become unfit for use assuming average frequency of use at The National Archives (Kew), i.e. 0.425 times per document per year, and with the threshold of one large missing piece per 100 sheets

DP	Years to unfit
300	55
400	160
500	450
600	1300
700	3700
800	10,000

CONCLUSIONS

Wear and tear, as a form of physical degradation, accumulates with use, and ultimately leads to objects becoming unfit for use. The principles described by reliability engineering (bathtub curve) were used to describe the process of accumulation of wear and tear, as a function of time. For the typical type of paper representative of ~70–80 % of Western library and archival collections, we looked at the 'random failure phase' when material weakens, as well as at the 'wear-out failure phase' during which reduction in DP becomes so pronounced that wear and tear (mechanical failure) becomes increasingly likely. If the frequency of object use is known, the curve can be experimentally determined.

In an experiment involving handling of mock archival and library items, we explored how wear and tear accumulates during handling challenges resembling reading. We conclude the following:

- The accumulation of wear and tear depends on DP for paper with DP < 800. For paper with higher DP, mechanical deterioration appears to accumulate at a very low rate, randomly and independently of DP. In our experiments, only 2–5 % of wear and tear accumulated on paper with DP > 300.

- For paper with DP < 300, at least one large missing piece accumulates per 100 sheets during each handling. This is accompanied by tearing and other, minor elements of mechanical deterioration. This value of DP could therefore be used as a safe threshold value for handling, if large missing pieces (containing text) are to be avoided. This is in agreement with previous folding test studies.

- Knowing the average frequency of use, it is possible to estimate the period of time a document with a given DP will remain fit for use if regularly handled. At the National Archives (Kew), wear and tear is not seen as an important issue for documents with DP > 600, where this period of time (400 years in average) is comparable to the long-term planning horizon (500 years).

From a collection management point of view, this research quantitatively confirms the important contribution of wear and tear to the loss of fitness for purpose, for documents of low DP. Taking the frequency of production into account, as well as the threshold number of large missing pieces per 100 sheets, the wear-out function could add several decades to the lifetime of even the most fragile objects (i.e. with DP < 300), while for objects with DP > 500, wear and tear on paper no longer represents a significant concern. Objects that are not handled do not reach the threshold fitness for use.

The research focused on mid-19th–mid-20th Century paper of Western origin: lignin containing, rosin sized, acidic, printed and non-coated. While our results show some variation in the accumulation of wear and tear for archival folders and bound objects, which in this research we treated as insignificant, it is possible that objects of significantly different size, weight or type of binding would behave substantially different.

The research quantitatively demonstrates how the desired lifetime of an object or a collection can be increased by adjusting the frequency of access, as mechanical deterioration is proportional to the number instances of user access. In the last paper in our series we will look at how DP decreases with time in relation to environmental parameters during storage (dose–response function) and to see how this could be visualised in ways that might be meaningful to collection managers and researchers alike.

AUTHORS' CONTRIBUTIONS

MS, CMG, CD, NB, KF, PB, EM, KN, WL, DT, GdB jointly developed the concept of this work. CMG and CD organised the experiment and developed guidance for assessment. CD, MS and CMG produced the samples. CMG, MS and CD analysed the data. All co-authors contributed to the manuscript. All authors read and approved the final manuscript.

ACKNOWLEDGEMENTS

The authors are grateful to colleagues who volunteered for the experiment: Tiphaine Bardon, Katherine Curran, Josep Grau, Andrej Kos, Julia Martinez, Keith Razey (in addition to some of the co-authors). We are also grateful to David Stevens of City Binders (London, UK) for his helpful advice.

COMPETING INTERESTS

The authors declare that they have no competing interests.

REFERENCES

1. Menart E, De Bruin G, Strlič M. Dose-response functions for historic paper. Polym Degrad Stab. 2011;96:2029–39.View Article

2. Rachwał B, Bratasz Ł, Łukomski M, Kozłowski R. Response of wood supports in panel paintings subjected to changing climate conditions. Strain. 2012;48:366–74.View Article

3. Rachwał B, Bratasz Ł, Krzemień L, Łukomski M, Kozłowski R. Fatigue damage of the gesso layer in panel paintings subjected to changing climate conditions. Strain. 2012;48:474–81.View Article

4. Klutke GA, Kiessler PC, Wortman MA. A critical look at the bathtub curve. IEEE Trans Reliab. 2003;52:125–9.View Article

5. Strlič M, Thickett D, Taylor J, Cassar M. Damage functions in heritage science. Stud Conserv. 2013;58:80–7.View Article

6. Walker G, Greenfield J, Fox J, Simonoff JS. The Yale Survey: a large-scale study of book deterioration in the Yale University Library. Coll Res Lib. 1985;46:111–32.View Article

7. Palm J, Cullhed P. Deteriorating Paper in Sweden, FoU-projektet för papperkonservering, Report No. 3. Stockholm: 1988.

8. Hunter D. Papermaking: history and technique of an ancient craft. New York: Dover Publ; 1987.

9. Buchanan S, Coleman S. Deterioration survey on the Stanford university libraries green library stack collection. Coll Res Lib. 1987;48:102–47.

10. Sobucki W, Srewniewska-Idziak B. Survey of the preservation status of the 19 and 20th century collections at the National Library in Warsaw. Restaurator. 2003;24:189–201.View Article

11. Strlič M, Grossi-Sampedro C, Dillon C, Bell N, Fouseki K, Brimblecombe P, Menart E, Ntanos K, Lindsay W, Thickett D, France F, De Bruin G. Damage function for historic paper. Part I: fitness for use. Herit Sci. 2015;3:33.View Article

12. Shroff DH, Stannett AW. A review of paper aging in power transformers. IEEE Proc. 1985;132:312–8.

13. Kolar J, Strlič M. Enhancing the lifespan of paper-based collections, AIC 35th annual meeting. Richmond: 2007.

14. Strlič M, Cséfalvayová L, Kolar J, Menart E, Kosek J, Barry C, Higgitt C, Cassar M. Non-destructive characterisation of iron gall ink drawings: not such a galling problem anymore. Talanta. 2010;81:412–7.View Article

15. Oriola M, Campo G, Ruiz-Recasens C, Pedragosa N, Strlič M. Conservation management scenario appraisal for painting canvases at Museu Nacional d'Art de Catalunya. Stud Conserv. 2015;60:S193–9.View Article

16. Zou X, Uesaka T, Gurnagul N. Prediction of paper permanence by accelerated aging. Part I: kinetic analysis of the aging process. Cellulose. 1996;3:243–67.View Article

17. Erhardt D, Tumosa CS, Mecklenburg MF. Material consequences of the aging of paper, 12th Preprints of the Triennial ICOM-CC meeting, vol. 2. London: James & James; 1999. p. 501–6.

18. SurveNIR website, http://www.science4heritage.org/survenir/collection. Accessed 8th July 2015.

19. Trafela T, Strlič M, Kolar J, Lichtblau DA, Anders M, Mencigar DP, Pihlar B. Nondestructive analysis and dating of historical paper based on IR spectroscopy and chemometric data evaluation. Anal Chem. 2007;79:6319–23.View Article

20. Kolar J, Malešič J, Kočar D, Strlič M, De Bruin G, Koleša D. Characterisation of paper containing iron gall ink using size exclusion chromatography. Polym Degrad Stab. 2012;97:2212–6.View Article

21. Dillon C, Lindsay W, Taylor J, Fouseki K, Bell N, Strlič M. Collections demography: stakeholders' views on the lifetime of collections. In: Ashley-Smith J, Burmester A, Eibl M, editors. Climate for collections conference, Munich, Doerner Institut, 7–9 November 2012, Postprints. London: Archetype; 2013. p. 45–58.

CHAPTER 11

Damage function for historic paper. Part III: Isochrones and demography of collections

Matija Strlič 1*, Carlota M. Grossi 1 , Catherine Dillon1, Nancy Bell2, Kalliopi Fouseki 1, Peter Brimblecombe3, Eva Menart1, Kostas Ntanos2, William Lindsay2, David Thickett4, Fenella France5 and Gerrit De Bruin 6

1Institute for Sustainable Heritage, University College London, London, UK.
2The National Archives, Kew, Richmond, Surrey, UK.
3University of East Anglia, Norwich, UK.
4English Heritage, London, UK.
5Library of Congress, Washington DC, USA.
6Nationaal Archief, The Hague, The Netherlands.

ABSTRACT

Background

In the context of evidence-based management of historic collections, a damage function combines aspects of material degradation, use, and consideration of material attributes that are important for satisfactory extraction of benefits from user interaction with heritage. In libraries and archives, it has been shown that users (readers and visitors) are mainly concerned with loss of textual information, which could lead to degradation being described as unacceptable, at which an object might become unfit for use and therefore damaged. The contribution explores the development of the damage function for historic paper based on data available in the literature.

Results

We have modelled the dose–response function taking into account 121 paper degradation experiments with known T, RH of the environment, and pH of paper. The function is based on the Arrhenius equation and published water absorption isotherm functions for paper. New isoperm plots have been calculated and isochrones have been developed. These are plots linking points of equal expected 'lifetime', i.e. time until an object is expected to reach the state of threshold fitness-for-use. We also modelled demographic curves for a well-characterised research collection of historic papers, exploring the loss of fitness for use with time.

Conclusions

The new tools enable us to evaluate scenarios of management of the storage environment as well as levels of access, for different types of library and archival paper. In addition, the costs and benefits of conservation interventions can be evaluated. The limitations of the function are the context of use (dark storage and reading), exclusive focus on the properties of an average paper type, and de-prioritised effect of pollutants; however, the latter can be considered separately. This work also demonstrates that transparent and publically accountable collection management decisions can be informed, and challenged by, effective interaction with a variety of stakeholders including the lay public.

KEYWORDS

Preventive conservation Collection modelling Fitness for use Wear and tear Libraries and archives

1. BACKGROUND

In our previous contributions in this series, we looked at what makes historic paper unfit for use in the context of general access in libraries and archives [1] and how mechanical degradation accumulates during handling of paper [2]. We have shown that readers and visitors are mostly concerned with loss of pieces of objects, especially if text is also missing. In such cases, degradation becomes unacceptable, objects are considered unfit for use and thus damaged. However, we have also shown that the level of acceptance of degradation depends on the value of the object [1].

Discoloration of paper (ink fading has not been explored) was not seen as important as loss of text, and accumulation of tears was also not seen as a significant contributor to loss of fitness by readers and visitors, either in the context of reading or in the context of exhibitions. Conversely, if objects were ascribed significant historical value, even loss of text did not make such objects unfit for use. This is of importance in the management of individual objects, however, in the context of conservation management of large library and archival collections, the value of individual objects may be difficult to take into account practically [1].

In daily interactions with objects, particularly in reading rooms, wear and tear may accumulate. We have shown [2] that this is of concern particularly for objects with degree of polymerisation (DP) of cellulose in paper between 300 and 800, while for objects with DP >800, wear and tear accumulates randomly. Objects with DP <300 are likely to develop significant wear and tear (a detached piece of paper with text) in a single instance of handling (reading).

Loss of fitness for use does not mean that information is no longer accessible from such objects; however, more resources are needed to do so: a conservation intervention or access under supervision might be required. It is reasonable to assume that objects (even those with low DP) that are not accessed do not accumulate wear and tear and may thus remain undamaged (i.e. without any text loss). In the context of average frequency of object use at The National Archives (Kew), large missing pieces of objects (such that contain text) may develop in ~55 years for paper with DP 300, ~160 years if its DP is 400 and ~450 years if it is 500 [2]. Thus, for objects with DP >500, such pieces in average develop over time intervals that are longer than the typical long-term planning horizon of 500 years [3], even for acidic papers. For objects with DP >800, the process becomes random [2].

We have thus explored two important value-based border criteria that are required to model the life of collections: (1) definition of fitness, and (2) acceptable planning horizon. To develop the damage function [3], it is now required to look into the dose–response function, linking loss of DP with time and variables that critically affect the rate of degradation of paper at the conditions of dark storage.

There is a substantial body of work on cellulose and paper degradation that has been summarised in recent reviews [4, 5] and books [6]. Much of this work is based on the earlier research into the kinetics of cellulose degradation by Ekenstam [7], Emsley and Stevens [8], Zou et al. [9] and others. We know that temperature (T), water content and the concentration of acids in cellulose critically affect the rate of cellulose degradation. A general equation can be developed based on the Ekenstam function [7]:

$$k \cdot t = \frac{1}{DP} - \frac{1}{DP_0},$$
(1)

where DP_0 and DP represent the degree of polymerisation of cellulose at time 0 and t, respectively, and k is the rate constant [year^{-1}]. Based on the work of Zou et al. [9], where

$$k = A_a e^{-\frac{E_a}{RT}}$$
(2)

And

$$A_a = A_{a0} + A_{a2} \cdot [H_2O] + A_{a5} \cdot [H_2O] \cdot [H^+],$$
(3)

Where

$$[H^+] = 10^{-pH}$$
(4)

and [H_2O] is water content in paper. It should be noted that pH of paper is an operationally defined quantity and not the pH of a true solution [10].

There has been considerable discussion in the literature about equilibrium moisture content in cellulose and paper. When developing isoperms, Sebera [11] presumed a linear dependence of water content on relative humidity in the environment, which might be a good approximation only in a limited range of RH values. This has later been addressed by Strang and Grattan [12] who proposed to use the Gavin-Anderson-de Boer equation, as explored for paper by Parker et al. [13]. IPI's eClimateNotebook® [14] also does not presume a linear dependence, but Strang and Grattan note "The IPI method, unfortunately, remains somewhat opaque, as the derivation has not been published". Water content of paper has also been expressed by Paltakari and Karlsson [15] using the following equation:

$$[H_2O] = \left(\frac{\ln{(1 - RH)}}{1.67 \cdot T - 285.655} \right)^{\frac{1}{2.491 - 0.012 \cdot T}}$$

(5)

where RH is relative humidity expressed as a ratio. Since this equation has been developed for 'fine paper' [15], it is worth noting that its use will result in a higher uncertainty of the developed model for papers with different fillers or sizing. Since it is expressed as a function of RH and T, the Paltakari and Karlsson function, plotted in Fig. 1, enables us to easily model A_{a0}, A_{a2} and A_{a5} in Eq. (3) on the basis of sufficient experimental data describing $k = f(T, RH, pH)$.

However, T, RH and pH are not the only variables contributing to paper degradation during dark storage. The effect of pollutants, specifically NO_2 and acetic acid, has recently been discussed by Menart et al. [16]. It has been shown that the effect of acetic acid is mostly insignificant in realistic experimental conditions, resembling storage conditions in post-industrial environments. On the other hand, the contribution of 10 ppb NO_2 to the rate of DP loss was comparable to that of ~4 °C for some papers (acidic and rag). NO_2 also contributed to significant yellowing of some types of paper; however, as we have shown this is not seen as an important element of fitness by general library and archival readers and visitors [1]. The contribution of O_3 would still need to be studied quantitatively, although its concentration is usually lower than that of NO_2 in archival and library repositories [16]. The contribution of biodeterioration to DP loss during dark storage is known to be substantial at RH >75 % in combination with unsuitable temperatures (>10 °C) [17], therefore, a dose–response function based on T, RH and pH should not be used for predictions at such high RH values.

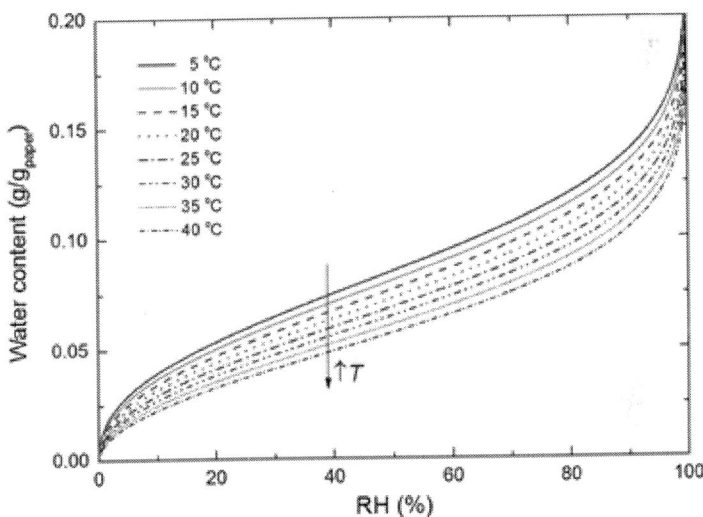

Figure 1 Paltakari and Karlsson water absorption isotherms for paper [15]. The equilibrium moisture content decreases with higher temperature ($\uparrow T$)

Taking the above constraints into account, we can now look at modelling of parameters in the dose response function for paper based on Eqs. (3) and (5), and on experimental data as published in the literature. This will lead to the development of improved isoperms. By combining the dose–response function with the value-based parameter of threshold fitness, we will develop the damage function for historic paper and look at how it can be applied in the context of management of archival and library collections.

2. METHODS

2.1 Experimental data

The experimental data were collected from 121 paper degradation experiments where T, RH and pH were provided, or degradation rates modelled at room temperature on the basis of DP and pH measurements of real objects. The following literature sources were used:

- Zou et al. [9] for 20 experimentally determined rates of degradation at 60, 70, 80, 90 and 100 °C and 2, 17, 58, 78 and 100 % RH for a bleached softwood bisulfite pulp and six bleached softwood kraft pulps, degraded as single sheets.

- Zou et al. [18] for 18 experimentally determined rates of degradation at room temperature for bleached kraft pulps, degraded as stacks.

- Baranski et al. [19], for 1 experimentally determined rate of degradation at 100 °C and 100 % RH for a bleached softwood pulp, degraded as a single sheet.

- Strlic et al. [6] for 73 experimentally determined rates of degradation at 60, 70, 80 and 90 °C and 65 % RH for cotton pulp, bleached sulfate pulp, Whatman filter paper No. 1 (Maidstone, UK) and historic papers from 1984 (50 % bleached sulfate hardwood pulp, 50 % bleached sulfite softwood pulp), 1870 (70 % cotton, 30 % wheat straw), 1938, (100 % sulfite softwood pulp) and office paper (70 % bleached sulfate softwood, 30 % bleached sulfate hardwood pulp), degraded as single sheets. Room temperature degradation rates for the real paper samples as calculated on the basis of the obtained Arrhenius models, were also used.

- Kolar and Strlic [20] for 9 rates of degradation historic papers made of bleached pulps at room temperature, modelled on the basis of measurements of pH and DP of papers of different age, degraded as stacks.

The above data has evidently been collected using different experimental approaches to accelerated degradation and it is likely that there are systematic differences in the methods of DP and pH determination. Furthermore, the above samples range from various types of bleached pulp to actual historic paper of different composition and manufacturing technology. It is likely that these characteristics (fibre type, beating, sizing, fillers, coatings etc.) contribute to the overall uncertainty of the developed dose–response function.

2.2 Modelling

STATA 14 was used for modelling of $\ln k$ using non-linear regression and the results are reported in Table 1, with the associated uncertainties. Isoperms and isochrones were plotted using OriginPro 9.0, while the demographic plots were calculated using Microsoft Excel.

Table 1 Values of parameters in Eq. (6)

Parameter	Estimate	Std. error	95 % Confidence interval	
			Lower bound	Upper bound
a_0	36.9812	1.6156	33.7827	40.1797
a_1	36.72	9.99	16.93	56.51
a_2	0.2443	0.0180	0.2087	0.2800
a_3	14299.8	632.8	13047.1	15552.5

2.2 Dose–response function and isoperms

The best fit was obtained using the following equation:

$$\ln(k) = a_0 + a_1 \cdot [H_2O] + a_2 \cdot \ln[H^+] - \frac{a_3}{T} \tag{6}$$

where the parameters are listed in Table 1. Based on a_3, the apparent activation energy is 118.9 kJ/mol, which is in line with values reported in the literature [6, 9, 11, 18]. It should be noted that Eq. (6) has not been derived directly from Eqs. (2) and (3), but is a linear combination of the factors that produced the best fit.

Combining Eqs. (4), (5) and (6), the dose response function for paper thus takes the following form

$$\ln(k) = 36.981 + 36.72 \cdot \left(\frac{\ln(1 - RH)}{1.67 \cdot T - 285.655} \right)^{\frac{1}{2.491 - 0.012 \cdot T}}$$

$$+ 0.244 \cdot \ln\left(10^{-pH}\right) - \frac{14300}{(T + 273.15)}, \tag{7}$$

where k is the rate constant (year^{-1}), RH is relative humidity expressed as a ratio, T is temperature (°C) and pH is that of paper.

Figure 2 shows the correlation of rate constants reported in the sources as indicated in "Experimental data", with the corresponding rate constants calculated using the available data on T, RH and pH, using the dose response function, Eq. (7).

Considering the diversity of data sources and experimental approaches, the agreement of modelled and measured rates (Fig. 2) is quite remarkable. The data points represent pure cotton linters sheets, cellulose pulps as well as naturally aged paper documents, and span a range of RH values from 2 to 100 %, temperatures from room to 100 °C, and pH values from 4 to 9. There is more substantial data scatter at lower temperatures (lower $-\ln k$ values), which is understandable, given the long experimental times and thus higher uncertainties.

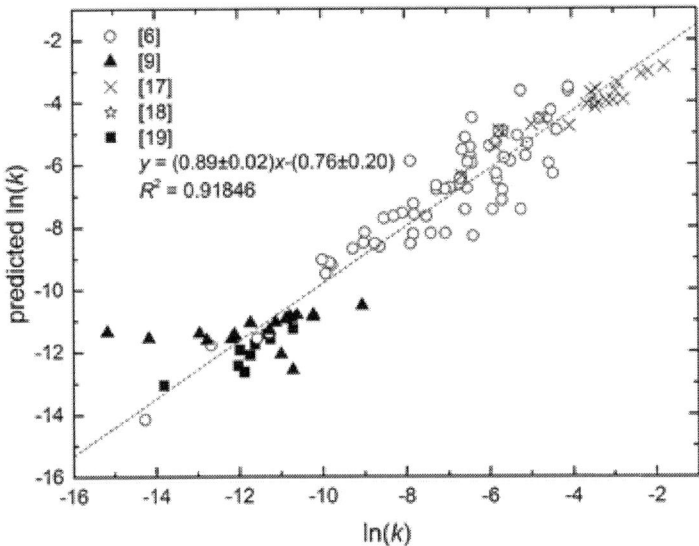

Figure 2 Comparison of the modelled rates of degradation using Eq. (7) and the observed rates of degradation as reported in the literature (as indicated)

In relation to measurement errors and uncertainties, it is useful to note that in the literature resources, DP was mostly determined using the cuprietylenediamine method, which, although standardised (ISO 5351/1:1981), is known to lead to potential overestimations of DP particularly of degraded cellulose [21], and in addition, there has been much recent research into the parameters of the Mark-Houwink-Sakurada equation on the basis of which DP is calculated from intrinsic viscosity. Most values in the cited literature were derived from the work of Evans and Wallis [22]. Additionally, since fairly large samples are required for DP determination, it is often the case that the amount of ash in historic samples is not determined due to the unavailability of sufficient sample amounts. Methods of pH determination of paper may also be quite different, although most show a reasonable correlation with the standard procedure [10].

Therefore, when attempting to validate Eq. (7) using new data obtained with substantially different analytical techniques, the above considerations should be taken into account.

The dose–response function can now be used to calculate new isoperm plots, linking combinations of values of variables T, RH and pH where equal permanence is expected, relative to the arbitrarily chosen set of 'standard' values of 20 °C and 50 % RH. However, the new dose–response function now depends not only on T and RH, but also on pH, and a number of isoperms can be plotted (Fig. 3).

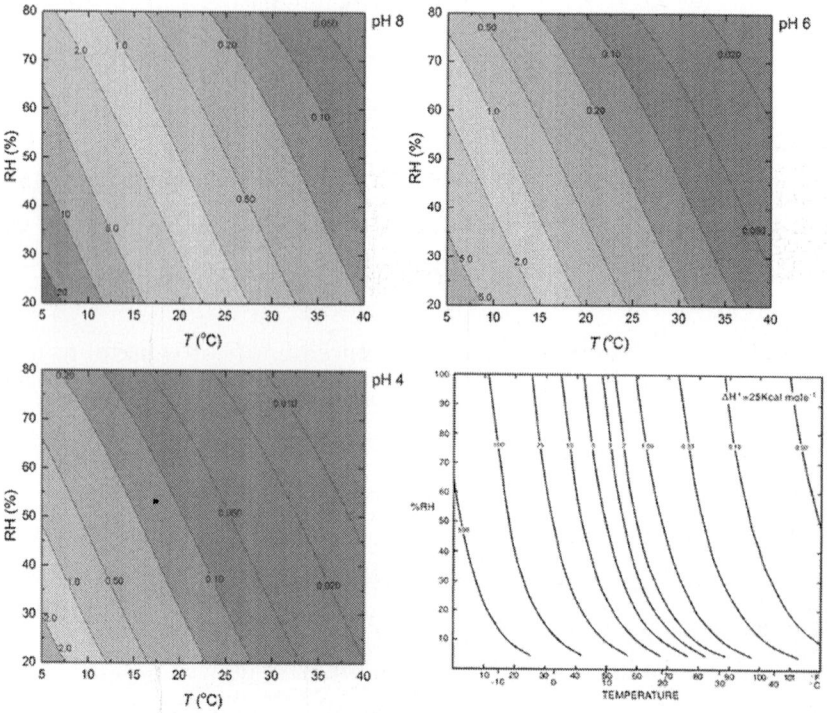

Figure 3 A comparison of the new isoperm plots for papers of pH 8, 6 and 4, with Sebera's isoperm plot [11]

In Fig. 3, we compare isoperms for alkaline paper (pH ~8), weakly acidic paper (pH ~6) and a strongly acidic paper (pH ~4) with Sebera's isoperm plot [11] which was based on acidic paper research. We now need to choose the arbitrary value of the 'standard' pH, in addition to Sebera's 'standard' 20 °C and 50 % RH, and it might be sensible to use print paper as the type of paper most prevalent in contemporary paper collections, with pH ~8. The colour scheme in Fig. 3 is chosen such that values above 1 tend towards amber and green, and values below 1 towards orange and red.

It is quite evident that the effect of moisture is more pronounced in our isotherm than in Sebera's work [11], where the effect of temperature was much more dominant. According to our isoperms, an increase of RH of ~20 % would lead to the same increase in the rate of degradation as an increase in T of about 4 °C.

Certainly, due to the lower overall stability of acidic paper, the isoperm values for progressively more acidic paper are lower. This can be more effectively explored in the set of isoperm plots in Fig. 4, where pH is not a fixed variable.

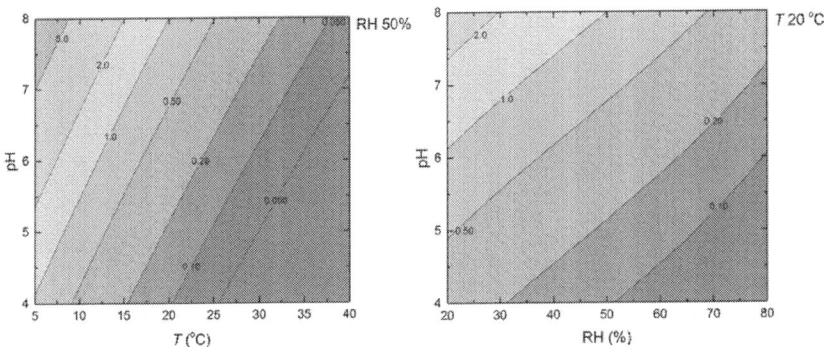

Figure 4 Isoperm plots for historic paper of different pH at fixed values of RH (50 %) and T (20 °C)

In Fig. 4, we can appreciate that a difference in pH of ~1.25 will have the same effect on durability as ~5 °C or 20 % RH. Temperature still has the dominant effect, while we see that the effect of pH and RH is very similar.

This has interesting implications in relation to interventive and preventive conservation. A change in pH of the material requires a conservation intervention (deacidification), which is common practice in paper conservation, though less so in mass scale. The isoperms present an argument for the use of deacidification, as a one-off investment into intervention will have the same effect as continuous storage of the same paper at ~10 °C cooler conditions in the long term (we return to this argument later with more details). However, if such cooler conditions could be achieved sustainably, e.g. in a passive storage building (external climate permitting), then it seems meaningful to attempt to do so.

This, in conjunction with the fact that the fairly usual fluctuations in T (±5 °C) and RH (±10 %) during storage in non-mechanically controlled environments (with even smaller fluctuations typical for storage boxes) do not seem to accelerate the degradation of paper [5], should motivate us to free ourselves from the constraints of rigid environmental management in paper-based collections, as was recently explored in the context of a large mechanically controlled archival repository [23] while simultaneously achieving significant preservation as well as economic benefits.

Isoperms are a useful tool to compare the preservation outcomes achievable in different environments, however, they do not allow us to visualise the expected remaining time an object might have until it becomes unfit. We will explore this in the next section.

2.3 Isochrones

In Part I of this paper series [1] we successfully defined that a large missing piece (containing text) is what makes a paper object unfit for use.

In Part II [2] we defined that the risk of this occurring during an instance of reading in the context of general access is on average very high (even 100 %) for objects with DP <300. We could thus define this value of DP as the threshold value at which objects are no longer suitable for general access.

With the dose–response function, Eq. (7), it is now possible to calculate the time required for an average object, with a certain starting DP_0 and pH, to reach this state. As with isoperms, we can calculate any number of combinations of T and RH during storage, where this expected period of time is equal—we call these lines isochrones. The concept was first used in the context of environmental management of collections of colour photographs [24], with the same purpose.

In Fig. 5, three sets of isochrones are calculated for three types of paper typical for archival and library collections: pH 5 and DP_0 of 600 (low-quality acidic paper from the first half of the 19th Century), pH 7 and DP_0 of 1500 (rag paper), and pH 8 and DP_0 of 2000 (contemporary print paper with $CaCO_3$ filler).

It can be appreciated that an average acidic paper may well survive the long-term planning horizon of 500 years if stored at an average annual temperature of 18 °C at 50 % RH or less. It needs to be stressed that this is based on what we consider to be an average acidic, rag or contemporary paper, and that individual collections may well reveal different and specific averages of pH and DP. However, this should be established in collection surveys which include measurements of pH and DP of individual items (discussed in "Demography of collections"). Based on Eq. (7), it would then be possible to calculate collection-specific isochrones.

It should also be stressed that at 10 ppb of NO_2 in the storage environment, this number is halved. If cooling is necessary to achieve this horizon, then a cost-benefit calculation can reveal if resources are better invested into energy continuously or into deacidification as a one-off investment.

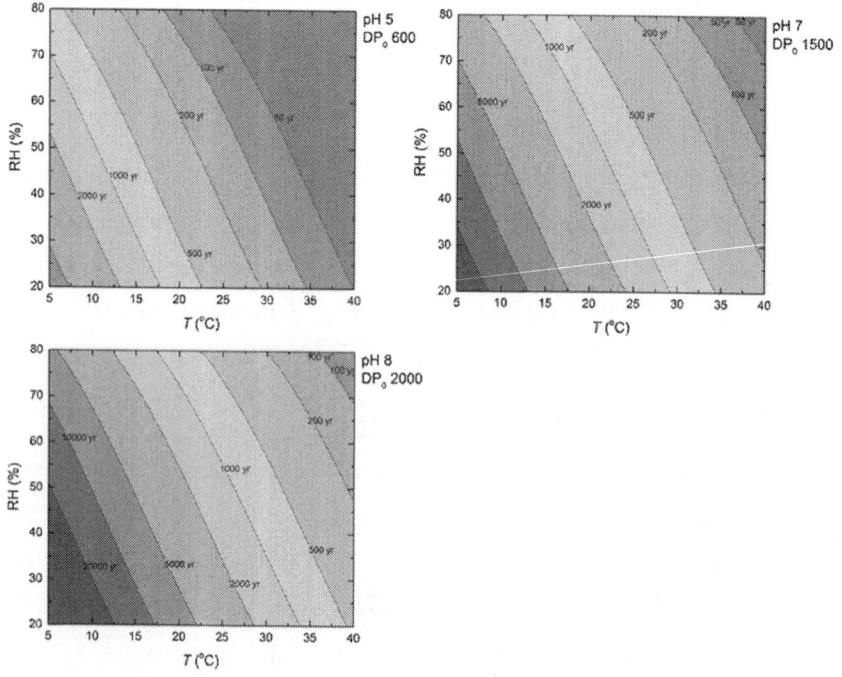

Figure 5 Isochrones for typical low-quality acidic paper from the first half of the 19th Century (pH 5, DP_0 600), typical rag paper (pH 7, DP_0 1500), and typical contemporary print paper (pH 8, DP_0 2000)

For rag paper and contemporary print paper, we see that there are few concerns, if any, with respect to their survival in temperate climates, even with a further reduction of the planning horizon due to the potential presence of pollutants.

It is worth stressing that these considerations are only valid for paper as the carrier of information. If the writing ink or pigment is particularly instable (e.g. iron-gall ink), then the isochrones for acidic paper could be used (pH 5, DP_0 600) as a first approximation, as it seems likely that acidity is one of the main contributors to iron gall ink degradation [25].

However, in Part II [2], we claimed that in case the frequency of access to objects is known, it is possible to calculate the expected lifetime expectancy of individual objects in a collection, if their current DP and pH are also known. Depending on the frequency of access and since missing pieces of text accumulate slowly, this will add a significant period of time to the overall expected lifetime, as explored in the Introduction for the National Archives (Kew). Such curves effectively represent demographic plots and we will explore these in the next section.

2.4 Demography of collections

Tools have become available using which it is possible to rapidly and non-destructively survey a representative sample of a large collection and determine the pH and DP of individual items [26, 27]. Using such data we can, on the basis of Eq. (7) and on the basis of the wear-out function [2], calculate the effects of both storage and frequency of access, on the time required for objects to become unfit for use.

In this work, we will use data collected on a model paper collection to demonstrate the benefits of doing so: the SurveNIR historic reference paper collection [28] has been extensively characterised and pH and DP data for 665 real historic papers are used here.

The SurveNIR collection does not represent a typical library collection in that it has only 32 % of acidic papers, 41 % of contemporary papers, 20 % of rag papers and 6 % of acidic papers that have previously been deacidified using the PaperSave process. In fact, we know that in a typical Western library or archival collection, the proportion of acidic papers is likely going to be 70–85 % [29], however, the data presented here may still be quite representative in comparative terms.

In Fig. 6, we explore four scenarios of collection management. Generally, we see that the collection as a whole will degrade in two waves: the first wave representing the acidic papers (which for most library and archival collections represents 70–85 % of the material [29], i.e. substantially more than the SurveNIR reference paper collection with ~32 %), and the second wave representing the rest. Very interestingly, we see that rag papers also degrade in two waves, which indicates that the more acidic ones behave similarly to the non-stable acidic paper.

Looking at the scenarios in more detail, we see that in case A, representing what might be current widespread practice, it is projected that ~2/3 of acidic paper will no longer be in a fit-for-use state in what is the long-term planning horizon acceptable to ~90 % library and archival users. About 1/3 or rag paper will also be in a similar state. Other types of papers will survive very much intact, in fact, we see that deacidified paper compares favourably with the more rapidly degrading acidic paper.

Case B represents a rather less strict fitness criterion, with every second page having some text missing—we see that the survival curve for acidic paper is slightly worse, with ~30 % still being fit after 500 years, while the other types of paper behave reasonably similarly to case A. This is because mechanical degradation accumulates at a significant rate only after paper degrades below DP 500 and as long as this is not the case, a stricter fitness criterion will have little effect.

Case C represents a library that is not accessed much—in average, each object is used only every 10 years. If so, then there are comparatively fewer instances when mechanical degradation can accumulate, and as a consequence, even degraded objects do not become unfit and only ~10 % of acidic paper will reach the fitness threshold, with the rest of the collection remaining fit for use.

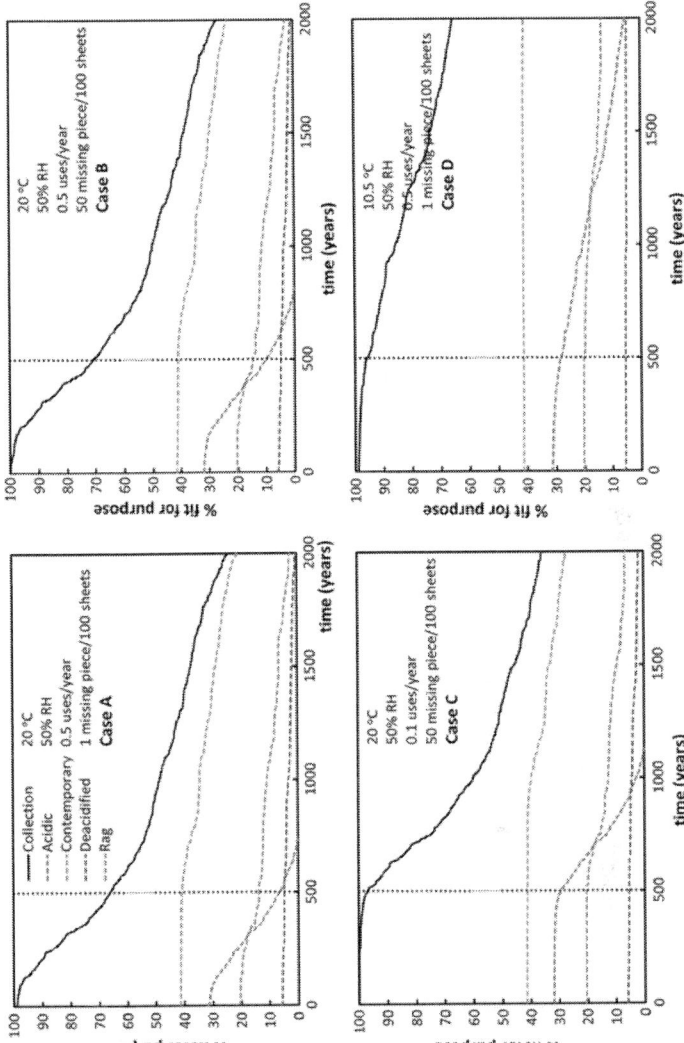

Figure 6 Demography of the SurveNIR reference paper collection at different conditions of storage and access, and different fitness criteria. **a** 20 °C, 50 % RH, used 0.5 times a year, fitness threshold: 1 missing piece with text per 100 pages; **b** 20 °C, 50 % RH, used 0.5 times a year, fitness threshold: 50 missing pieces with text per 100 pages; **c** 20 °C, 50 % RH, used 0.1 times a year, fitness threshold: 50 missing pieces with text per 100 pages; **d** 10.5 °C, 50 % RH, used 0.5 times a year, fitness threshold: 1 missing piece with text per 100 pages. The *vertical dashed line* represents the long-term planning horizon of 500 years

Case D represents a preventive conservation scenario where the environment is kept at the same average annual temperature as the outdoor climate in London (10.5 °C), while the other variables remain the same. As expected, in comparison to Case A, all the objects will survive for longer and because even acidic paper will chemically degrade more slowly, only 10 % of it will reach the fitness threshold if used once every 2 years.

From a collection management perspective, we thus have evidence to consider several options: more favourable storage environment, less access, or conservation intervention, and depending on the institutional policy in question, a suitable balance can be struck "between the often conflicting demands of the care of the collection, the use (…) of collection items and energy economy", as required by a recent environmental guideline [30].

2.5 The public as a stakeholder

Throughout this series of papers, we considered the engaged public (readers, visitors) as a stakeholder in the decision-making process, although it may not be entirely evident why this should be the case. There are certainly cases where the stakeholders may need to have a certain level of understanding of conservation to define when the risk of unacceptable damage to an item is too high, as was the case of the threshold strength for painting canvases, where it is not reasonable to expect that members of the general public have any expectation of what strength is required for a canvas to be safely transported or restored [31].

However, where fitness can be determined empirically and it negatively affects the way that users extract benefits from an object, then fitness thresholds can be determined collaboratively with the public. This has been done in relation to fading of colour photographs [24] where it has also been shown that there is no statistically significant difference in what general archival readers and 'expert' users consider to be damage. In the first paper in this series we also discussed that there is little difference in the expectations with respect to the future use of collections between users as members of the public, and expert users, i.e. curators and researchers [1].

However, evidence from lay users should not only be used to legitimise decision making. Evidence from lay users in this project was more complex and had the potential to challenge expert opinion. For example, desired lifetime was shown to be much shorter on average for historic house contexts than archives. Desired lifetime was also influenced by characteristics of documents such as their age. In addition, discolouration was found to be less important to fitness-for-use than tears and missing pieces. Finally, profiling revealed different types of users, with some valuing broad social and personal aspects of their interaction with documents and some focused solely on deriving information content from documents with less regard for the survival of documents.

It would appear that the public's perception of risk and change may be much less conservative than the expert view. Lay users also appear to be open to biases relating to features of documents and context. Indeed there is not one public perception of damage and value but diverse perceptions in different circumstances. Decision makers may understandably have questions about the validity and objectivity of evidence gained through user engagement and how to use the evidence. It may be perceived that user engagement generates more questions and encourages rather than resolves conflict [32, 33].

Expecting a consensus view from user engagement exercises about the parameters of collections management would be unrealistic. In addition it would be unethical and compromise research quality to divest decision making responsibility from experts to lay users, who generally lack research expertise and do not have the necessary technical knowledge. At the same time the decision maker may also be concerned that experts' views can be incomplete, conflict with each other, be uncertain or not be grounded in end-users' concerns [34]. An appropriate role for evidence from lay users' needs to be specified.

User engagement can be one part of the decision maker's effort to fully consult with all relevant stakeholders to ensure a holistic picture of the consequences of a collection management decision. Lay users, and those who represent them, can give information about what matters to them, how they will be affected in their work or daily life and the relative importance they place on different aspects of change. These insights have the potential to inform or challenge expert definitions of damage, value and planning horizons. The evidence from users can also highlight potential future controversy and make explicit the social consequences of decisions. Therefore, it can be argued that user engagement has a role to play in technical and knowledge-based collections decisions by helping to make expert decisions transparent and by focusing priorities for decision making on the public's values. Hence, user engagement has the potential to ensure that collections management decisions can be related back to outcomes for people and help an organisation demonstrate their social and economic impact.

Creating a dynamic relationship between stakeholder groups, users, policy makers, collection managers, offers an opportunity to effect change, not just measure it, thus improving the context of decision making. Stakeholders' views contribute a distinct body of data to inform how resources are allocated and why, increasingly important for public accountability and transparency.

3. CONCLUSIONS

We have shown that in order to develop a damage function, the following elements are required:

- Dose response function, linking degradation rate with environmental variables and material properties

- Wear-out function, linking chemical degradation with accumulation of mechanical degradation due to use

- Fitness-for-use threshold defined as the state of an object where its use (e.g. reading, or display) is no longer satisfactory and leads to significantly reduced benefits.

To develop the above elements, we used degradation rate data as published in the literature, and worked with users of collections (visitors and readers) to develop an understanding of what could be considered as damage, leading to loss of fitness of an object.

The damage function for historic paper enabled us to calculate new isoperm plots, linking points of equal expected permanence. We also developed isochrones, plots linking points of equal expected 'lifetime', i.e. time until an object is expected to reach the state of threshold fitness for use.

The function, in conjunction with the long-term collection management planning horizon, enables us to evaluate scenarios of management of the storage environment as well as levels of access, for different types of library and archival paper. If the pH and DP data for a representative sample of a large collection is known, demographic curves can be plotted, looking at progressive loss of fitness for use of collection items. These enable us to develop collection-specific environmental and access management scenarios. Equally, the model enables us to evaluate the costs and benefits of conservation interventions, in relation to the cost of preventive conservation.

We would like to stress that the model has been developed for what could be considered to be the majority of archival and library paper, i.e. either gelatine or rosin sized; however, the model does not take into account coated, transparent or other speciality papers, and neither does it take into account the effect of inks. More detailed explorations of the dose-response function might be necessary for specific paper types.

The damage function is specific to the context of use, i.e. reading and dark storage, because the degradation variables have been prioritised accordingly: e.g. light is not prioritised, and equally, pollutants are not seen as a significant contribution to the overall rate of degradation because (1) their concentrations in post-industrial environments are small, and (2) they may only significantly contribute to degradation of certain types of paper. Additionally, discolouration is not seen as a major parameter of fitness in the context of library and archival materials (although it might be relevant in other contexts, e.g. in art collections).

It is important to note that objects that have reached the unfit state do not cease being useful: it is only that due to their fragile state, more resources are required to enable access to information, e.g. supervised access or access in a digital format.

Finally, it may be useful to stress that the planning horizon is not a fixed point in the future but merely a planning tool to enable collection management decisions to be made at a given point in time. When and if the next generation of users is asked about what they consider to be a suitable horizon, about what they consider as unacceptable degradation and about whether and how they value paper-based information, we might get different responses—but this is for the next generation of collection managers to be concerned about.

AUTHORS' CONTRIBUTIONS

MS, CMG, CD, NB, KF, PB, EM, KN, WL, DT, GdB jointly developed the concept of this work. CMG and MS developed the computational model and all co-authors contributed to interpretation. CMG analysed the data. All co-authors contributed to the manuscript. All authors read and approved the final manuscript.

COMPETING INTERESTS

The authors declare that they have no competing interests.

REFERENCES

1. Strlič M, Grossi-Sampedro C, Dillon C, Bell N, Fouseki K, Brimblecombe P, Menart E, Ntanos K, Lindsay W, Thickett D, France F, De Bruin G. Damage function for historic paper. Part I: fitness for use. Her Sci. 2015;3:33.

2. Strlič M, Grossi-Sampedro C, Dillon C, Bell N, Fouseki K, Brimblecombe P, Menart E, Ntanos K, Lindsay W, Thickett D, France F, De Bruin G. Damage function for historic paper. Part II: wear and tear. Her Sci. 2015;3:36.

3. Strlič M, Thickett D, Taylor J, Cassar M. Damage functions in heritage science. Stud Conserv. 2013;58:80–7.

4. Zervos S. Natural and accelerated ageing of cellulose and paper: a literature review. In: Lejeune A, Deprez T, editors. Cellulose structure and properties, derivatives and industrial uses. Happauge NY: Nova Science Publishers; 2010. p. 155–203.

5. Menart E, De Bruin G, Strlič M. Dose-response functions for historic paper. Polym Degrad Stab. 2011;96:2029–39.

6. Strlic M, Kolar J. Ageing and stabilisation of paper. Ljubljana: National and University Library; 2005.

7. Ekenstam A: Über das Verhalten der Cellulose in Mineralsäure-lösungen, II. Mitteil: Kinetisches Studium des abbaus der Cellulose in Säure-lösungen. Ber Dtsch Chem Ges. 1936; 69:553–559.

8. Emsley AM, Stevens GC. Kinetics and mechanisms of the low-temperature degradation of cellulose. Cellulose. 1994;1:26–56.

9. Zou X, Uesaka T, Gurnagul N. Prediction of paper permanence by accelerated aging I. Kinetic analysis of the aging process. Cellulose. 1996;3:243–67.

10. Strlič M, Kolar J, Kočar D, Drnovsek T, Selih VS, Susic R, Pihlar B. What is the pH of alkaline paper? e-Preservation Sci. 2004;1:35–47.

11. Sebera DK. Isoperms. An environmental management tool, http://cool-palimpsest.stanford.edu/byauth/sebera/isoperm/index.html. Accessed 3 Aug 2015.

12. Strang T, Grattan D. Temperature and humidity considerations for the preservation of organic collections—the isoperm revisited. e-Preservation Sci. 2009;6:122–8.

13. Parker ME, Bronlund JE, Mawson AJ. Moisture sorption isotherms for paper and paperboard in food chain conditions. Packag Technol Sci. 2006;19:193–209.

14. eClimateNotebook: https://www.imagepermanenceinstitute.org/environmental-management/eclimatenotebook. Accessed 3 Aug 2015.

15. Paltakari JT, Karlsson MA: Determination of specific heat for dry fibre material. 82nd Annual meeting, Technical Section, CPPA, 1996, B117-B120.

16. Menart E, De Bruin G, Strlič M. Effects of NO_2 and acetic acid on the stability of historic paper. Cellulose. 2014;21:3701–13.

17. Sedlbauer K. Prediction of mould growth by hygrothermal calculation. J Build Phys. 2002;25:321–36.

18. Zou X, Uesaka T, Gurnagul N. Prediction of paper permanence by accelerated aging. 2. Comparison of the predictions with natural aging results. Cellulose. 1996;3:269–79.

19. Baranski A, Lagan JM, Lojewski T. Acid-catalysed degradation. In: Strlic M, Kolar J, editors. Ageing and stabilisation of paper. Ljubljana: National and University Library; 2005. p. 85–100.

20. Kolar J, Strlič M: Enhancing the lifespan of paper-based collections. In: 35th AIC Annual Meeting Richmond VA, 16–20 April 2007.

21. Strlič M, Kolar J, Zigon M, Pihlar B. Evaluation of size-exclusion chromatography and viscometry for the determination of molecular masses of oxidised cellulose. J Chromatogr A. 1998;805:93–9.

22. Evans R, Wallis A. Comparison of cellulose molecular weights determined by high performance size exclusion chromatography and viscometry. Proc Fourth Int Symp Wood Pulping Chem Paris. 1987;1:201–5.

23. Hong SH, Strlič M, Ridley I, Ntanos K, Bell N, Cassar M. Climate change mitigation strategies for mechanically controlled repositories: the case of The National Archives, Kew. Atm Env. 2012;49:163–70.

24. Fenech A, Dillon C, Ntanos K, Bell N, Barrett M, Strlič M. Modelling the lifetime of colour photographs in archival collections. Stud Conserv. 2013;58:107–16.

25. Kolar J, Štolfa A, Strlič M, Pompe M, Pihlar B, Budnar M, Simčič J, Reissland B. Historical iron gall ink containing documents—

properties affecting their condition. Anal Chim Acta. 2006;555:167–74.

26. Lichtblau DA, Strlič M, Trafela T, Kolar J, Anders M. Determination of mechanical properties of historical paper based on NIR spectroscopy and chemometrics—a new instrument. Appl Phys A. 2008;92:191–5.

27. Trafela T, Strlič M, Kolar J, Lichtblau DA, Anders M, Pucko Mencigar D, Pihlar B. Non-destructive analysis and dating of historical paper based on IR spectroscopy and chemometric data evaluation. Anal Chem. 2007;79:6319–23.

28. SurveNIR website, http://www.science4heritage.org/survenir/collection (accessed 3 August 2015).

29. Barański A, Konieczna-Molenda A, Łagan JM, Proniewicz LM. Catastrophic room temperature degradation of cotton cellulose. Restaurator. 2003;24:36–45.

30. PAS198:2012. Specification for managing environmental conditions for cultural collections. The British Standards Institution, 2012.

31. Oriola M, Campo G, Ruiz-Recasens C, Pedragosa N, Strlič M. Conservation management scenario appraisal for painting canvases at Museu Nacional d'Art de Catalunya. Stud Conserv. 2015;60:S193–9.

32. Fouseki K, Sakka N. Valuing an ancient palaestra in the centre of Athens: The public, the experts, and Aristotle. Conserv Manag Archaeol Sites. 2013;15:30–44.

33. McDonald H. Understanding the antecedents to public interest and engagement with heritage. Eur J Marketing. 2011;45:780–804.

34. Fouseki K. Community voices, curatorial choices: community consultation for the 1807 exhibitions. Mus Soc. 2010;8:180–92.

CHAPTER 12

Users' Perception of University Library Resources and Services in South East Zone of Nigeria

Obiozor-Ekeze Roseline Nkechi
Nnamdi Azikiwe University Library, Awka, Nigeria

ABSTRACT

This study seeks to determine how users of the university libraries perceive the services rendered to them. Department of Mechanical Engineering was chosen as the population of this study. It is heavily populated with students seeking for information. 600 copies of questionnaire were made, only 450 copies were returned, well-filled, that is 75% of the questionnaire. Results were gathered and findings were made. The result showed that University Libraries are underfunded, they lack qualified personnel and also the university libraries do not have a general standard that guild them to access if actually users of the libraries are satisfied with the library or not.

KEYWORDS

University Libraries, Information Services, Department of Mechanical Engineering, Students, Standards, South East of Nigeria

1. INTRODUCTION

There are about ten (10) universities in the South East of Nigeria and each of them has not less than eight (8) faculties. South East Zone of Nigeria is heavily populated with students aiming to acquire one specialization or the other. It is really an information seeking zone in Nigeria . The universities have libraries that support them to achieve their objectives: be supportive to teaching, learning and research, be supportive to the mission and goals of their institutions. Gyure (2008) stated that university library is "the heart of an academic institution for accumulating knowledge. Some years back, the university libraries were acquiring materials on printed forms only, but presently, it is a different situation. The application of new technologies have changed and reshaped the needs, perception and expectation of users. Library's users expect more from the libraries. Since the advent of internet in libraries, the university libraries satisfy their users more. Internet resources have helped the collections of the university libraries to be rich. Users can access the online public access catalogue (OPACS) for more information unlike when it was only the libraries traditional catalogues. In a research carried out by Baikady & Mudhot (2011) on web learning resources in a medical college in India, they discovered that staff and students prefer using web resources than going to the library to use books or journal, which they are not sure are in the library. They use of OPAC is one of the greatest things that have happened to the university libraries. It provides current information resources to the users.

University libraries provide many services to their users: Collection development, reference services, organization of knowledge, users' education etc. These services are being carried out by the Librarians. Murphy (2000) , Beverly, Both and Bath (2003) reported the unique roles played by the university libraries through their librarians, as information professionals, they gather, organize and coordinate access to the best available information resources for library users. During collection development, the librarians liaise with the faculty representatives so as to acquire the necessary materials. For reference services, the users of the libraries are provided with quality information through the librarians. Bureau Adelabu (2004) calls these librarians "information providers". He reported that librarians also provide online chart and email for the users. University library organizes orientation courses for the newly admitted students as information literacy. It takes the form of a lecture, guided tours and formal bibliographic instruction. Glynn and Wu (2003) pointed out that the university libraries offer training ser- vices to the students and researchers in finding information effectively on the web for academic purposes.

What librarians are thinking of now is how to provide quality services to their clienteles. In developed countries, university libraries have standards and policy that guide them to evaluate services rendered to users of libraries. Cullen and Calvert (1993) observe that several ways could be adopted to evaluate library resources and services. He noted that input evaluation based on finances, staff resources and collection and output evaluation based on process efficiency measures are indices of how services are provided by the library.

In South East Nigeria, university libraries have a lot of problems hindering them from satisfying their users especially in science and technology. However, are studies on individual universities from other geopolitical zone on users' satisfaction, but the researcher carried out this study collectively on user's perception of libraries services and resources under Mechanical Engineering Departments, in South East Universities in Nigeria.

2. PURPOSE OF STUDIES

The purpose is to find out if staff and students of Mechanical Engineering using the university libraries are satisfied with the services and resources provided in the university libraries. Specifically to know if

1. The university library provides current books and journals for the department.

2. Know if internet service provided in the library is functional.

3. Know if there is a science based librarian in the university library that assist in providing services for the department.

3. RESEARCH QUESTIONS

1. What is the state of collection development for your department in your university library?

2. Is the internet service provided functional?

3. How capable are the librarians in your university libraries in handling science and technology matters?

4. METHODOLOGY

This study adopted descriptive survey design. The area of study is the university libraries in South Eastern Zone of Nigeria. The population of study is only registered staff and 300 level students from departments of Mechanical Engineering. This department was chosen because it is a common department in all the chosen universities and also heavily populated with male students. Their ages are between 19 and 24 years. The staffs chosen were mainly younger lecturers that have not worked for 5 years.

5. INSTRUMENT

A questionnaire on "Users Perceptions on University Library Resources and Services" was constructed for data collection. 600 copies of the questionnaire were duplicated and distributed with research assistances. They were collected after two (2) weeks. 450 are the respondents. Their copies of questionnaires were well completed were returned. That is 75% of the questionnaires.

6. FINDINGS

It was gathered that 60% of the respondents reported that the university libraries do not have current materials on Science and Technology. Current foreign journals are not in the libraries and the local journals are not continuous. The University libraries are under funded, especially the state and private university libraries. Federal university libraries have more advantages over the state and private university libraries. Agboola (2000) reported that for about five (5) decades federal and state governments have been the sole financers of libraries in Nigeria. In 1993, the National Universities Commission (NUC) legislated that 10% of the approved recurrent budget for Nigeria Federal Universities be set aside for library development. But the goodwill was withdrawn in 2001. The government is not actually encouraging the growth of libraries in Nigeria. Under funding affects the growth of libraries greatly. In collection development, underfunding affects the acquisition of materials, especially in Engineering and Technology. Meaningful materials cannot be purchased without much fund. Majority of university libraries in Nigeria are not internet connected because of lack of fund. Where it exists, it may not be functional. For users satisfaction to be achieved in university libraries, both printed and electronic resources must be availa- ble for consumption.

Another hindrance is the issue of not having the qualified personnels to work in science and technology libraries. Professional librarians do not have subject specialization. Subject background helps librarians to serve better in information services. Watson (2005) explains that if a non-science librarian for instance should manage a science library, the library will be put in a tragic damage. Both the collection development and reference services will be in a mess. To support this issue on subject background in librarianship, Ifidon (2005) remarked that subject specialization helps the librarians in the area of which he specializes in, enables the librarian to have a remote access to any information within his subject area. Another point gathered, was the issue of not having standard or policies that guild them in evaluating their services.

From the look of things, each of these universities libraries is operating on its own, according to the objectives of its parent institution. National Universities Commission stipulated a standard but it is not being used by the university libraries in Nigeria.

7. CONCLUSIONS

University libraries in South East Zone of Nigeria should wake up to what is happening around the world. There is no way the libraries can boost of information satisfaction to their users without having information resources and qualified personals to render the services.

The printed books and journals are not sufficient for users in mechanical engineering. The universities are over populated with students and researchers lacking sufficient information. Only the privileged students use their computers to search for information or they use the Cybercafés. Federal and state governments should increase the allocations given to universities so as to support their libraries.

Another point is giving support to other staff in libraries to study librarianship. Also Librarians working in the university libraries should have subject specializations especially in sciences to support them in services and also aquire ICT trainings. Some librarians in some areas should be sent for specialist courses. When university libraries provide quality information services to researchers, quality results will be gotten. This will contribute to the economic development of Nigeria .

REFERENCES

1. Agboola, T. (2000). Five Decades of Nigerian University Libraries. A Review of Library, 50, 280-287.

2. Baikady, M., & Mudhot, M. V. (2011). Web as a Learning Resource at the Medical College Libraries in Coastal Karnataka: Perception of Faculty and Students. DESIDOC Journal of Library & Information Technology, 31, 121-135.

3. Beverly, Both, & Beverly, Bath (2003). Subject Specialist Librarians in Higher Education: A Selective Review of the Literature, with a Brief Postscript Relating to Middlesex Polytechnic Library. Learning Resources Bulletin, 4, 22-31.

4. Bureau Adelabu (2004). Professional Staff of Tomorrow's Future in African University Libraries Some Poslates and Proposals. International library Review, 6, 299-308.

5. Cullen, R. J., & Calvert, P. J. (1993). Further Dimensions of Public Library Effectiveness: Report on a Parallel New Zealand Study. Library and Information Science Research, 5, 143-164.

6. Glynn, T., & Wu, C. (2003). New Roles and Opportunities for Academic Library Liaisons: A Survey and Recommendations. Reference Services Review, 31, 122-128.

7. Gyure, D. (2008). The Heart of the University. Winterthur Portfolio, 42, 107-132.

8. Ifidon, S. E. (2005). Information Availability and the Extent of Use in Public Library. Archival and Information Science Journal, 7, 65-74.

9. Murphy, C. (2000). The Role of the Subject Specialist in British and American University Libraries: A Comparative Study. Libri., 29, 1-19.

10. Watson, E. M. (2005). Subject Knowledge in the Health Sciences Library: An Online Survey of Canadian Academic Health Sciences Librarians. Journal of Medical Library Association, 93, 459-466.

CHAPTER 13

Use of citation analysis to predict the outcome of the 2001 Research Assessment Exercise for Unit of Assessment (UoA) 61: Library and Information Management

Alison Holmes and Charles Oppenheim

Department of Information Science Loughborough University ,Loughborough

ABSTRACT

A citation study was carried out to predict the outcome of the Higher Education Funding Councils' 2001 Research Assessment Exercise (RAE). The correlation between scores achieved by academic departments in the UK in the 1996 Research Assessment Exercise, and the number of citations received by academics in those departments for articles published in the period 1994-2000, using the Institute for Scientific Information's citation databases, was assessed. A citation study was carried out on all three hundred and thirty eight academics who teach in the UK library and information science schools. These authors between them received two thousand three hundred and one citations for articles they had published between 1994 and the present. The results were ranked by Department, and compared to the ratings awarded to the departments in the 1996 RAE. On the assumption that RAE scores and citation counts are correlated, predictions were made for the likely scores in the 2001 RAE. Comments were also made on the impact of staff movements from one Higher Education Institution to another.

1. INTRODUCTION

According to Popper's well-established philosophy of science, a discipline can only be considered a "science" if it makes falsifiable predictions, i.e., the theories underlying the discipline can be applied to new and untested situations and a prediction made of the likely outcome. Although the term "information science" is well established, under these criteria there are few, if any theories in information science outside of the field of modelling and system prediction in the information retrieval area outside of the field of modelling and system prediction in the information retrieval area that allow researchers to predict the outcome of a particular experiment (Summers, *et al.*, 1999). For example, the well-known inverse relationship between Recall and Precision does not allow one to predict what the precision or recall of a particular retrieval experiment will be, and the Bradford-Zipf "law" does not allow one to predict the size of a core corpus of literature in a particular subject. A cursory examination of the information science literature demonstrates the rarity of the words "predict", "predicting", "predicted" or "prediction" in the titles of papers. If one follows the Popperian approach, the second word in the term "information science" appears to be unjustified. In one area of information science research other than information retrieval, namely citation analysis, however, predictions have been made (Oppenheim, 1979). The work described here follows in that tradition.

2. THE RAE

The Higher Education Funding Councils in the UK carry out a Research Assessment Exercise (RAE) every few years (the last one was in 1996) to grade the research output of University Departments in the UK. The next Exercise will take place in 2001. The RAE is conducted jointly by the Higher Education Funding Council for England (HEFCE), the Scottish Higher Education Funding Council (SHEFC), the Higher Education Funding Council for Wales (HEFCW) and the Department of Education for Northern Ireland (DENI).

The nominal purpose of the RAE is to enable the higher education funding bodies to distribute some of their public funds to Universities selectively on the basis of the quality of research carried out in each Department. The RAE assesses the quality of research in Higher Education Institutions (HEIs) in the UK. Around £5 billion of research funds per annum will be distributed in response to the results of the 2001 RAE (HEFCE, 1994). The RAE has become one of the most important features of UK academic life, and in practice, the results are widely used for promotional purposes.

The RAE provides quality ratings for research across all disciplines. Panels use a standard scale to award a rating for each submission. Ratings range (HEFCE, 1995) from 1 to 5*:

	... submitted showing some evidence of international excellence.
3a	Quality that equals contribution to levels of international excellence in over two-thirds of the research activity submitted, possibly showing evidence of international excellence.
3b	Quality that equals contribution to levels of international excellence in more than half of the research activity submitted.
2	Quality that equals contribution to levels of international...

Panel members are appointed from the list of nominations of experienced and well-regarded members of the research community, and users of research. The inclusion of user representatives on certain panels is intended to enhance the assessment process by making available a user perspective. Panels use their professional judgement to form a view about the overall quality of research activity described in each submission in the round, taking into account all the evidence presented.

Institutions have been invited to make submissions containing information on staff in post on the census date, 31 March 2001. These submissions include a selection of publications emanating from the Department in question during the assessment period (1 January 1994 to 31December 2000, for Arts and Humanities subjects, and 1 January 1996 to 31 Decemb er 2000 for other subjects).

The outcomes of the assessment will be published in December 2001, and will provide public information on the quality of research in universities and higher education colleges throughout the UK. These results will show the number and proportion of staff submitted for assessment in each case, and the rating awarded to each submission. The funding bodies will also produce:

a) published report by each panel confirming their working methods and giving a brief account of their observations about the strengths, weaknesses and intensity of activity of the research areas falling within the Unit of Assessment (library and information management is UoA 61);

b) feedback on each submission summarising the reason for the rating awarded with reference to the panel's published criteria.

Those parts of submissions that contain factual data and textual information about the research environment will be published on the Internet. This will include the names of selected staff and their research publication output.

A large amount of public money is spent on the RAE. It involves considerable opportunity costs as well as actual financial expense by both the HEIs concerned, and the panels of experts in their particular subjects.

3. THE RAE AND CITATION COUNTS

Many studies have demonstrated that there is a strong correlation between citation counts and ratings of academic excellence as measured by other means, for example receipt of honours, receipt of research funding, editorships of major journals and peer group ratings. The subject has been discussed in some depth by Baird and Oppenheim (1994), Garfield (1979) and Cole and Cole (1973).

Until now, only three studies have been carried out regarding the correlation between RAE counts and citations. In the first study, Oppenheim (1995) demonstrated the correlation between citation counts and the 1992 RAE Ratings for British library and information science University Departments. He concluded that the cost and effort of the RAE might not be justified when a simpler and cheaper alternative, namely a citation counting exercise, could be undertaken.

Seng and Willett (1995) reported a citation analysis of the 1989 and 1990 publications of seven UK library schools. The total number of citations, the mean number of citations per member of staff, and the mean number of citations per publication were all strongly correlated, with the ratings that these departments had achieved in the 1992 RAE. An analysis of the citedness of different types of research output from these departments was also carried out.

Oppenheim (1997) conducted the third study that showed the correlation between citation counts and the 1992 RAE ratings for British research in genetics, anatomy and in archaeology. He found that in all three cases, there was a statistically significant correlation between the total number of citations received, or the average number of citations per member of staff, and the RAE score.

Recent reviews on citation counting can be found in Baird and Oppenheim (1994) and Liu (1994). An overview of aspects of research evaluation is provided by Hemlin (1996), who pointed out that there is a strong positive correlation between citation counts and other means of measuring research excellence, despite justified concerns about the validity of citation counting.

The correlation of RAE ratings with other measures has not been studied much. Colman *et al.*, (1995) carried out an evaluation of the research performance of British University politics departments. The research performance of forty-one British university politics departments was evaluated through an analysis of articles published between 1987 and 1992 in nine European politics journals with the highest citation impact factors. Annual performance scores were obtained by dividing each departments number of publications in these journals in each year (departmental productivity) by the corresponding departmental size. These scores were summed to obtain a research performance score for each department over the period of assessment. They correlate significantly with research performance scores from two previous studies using different methodologies: Crewe's per capita simple publication count for the years 1978 — 1984; and the Universities Funding Council's research selectivity ratings covering the years 1989 - 1992.

The correlation between a 1985 University Grants Committee research ranking exercise and a so-called 'influence weight' measure of journals in which the academics publish was considered by Carpenter *et al* (1988).

Citation studies have been used to compare different departments within one country in the same subject area (Winclawska, 1996), and of different departments around the world (Nederhof & Noynes, 1992).

Ajibade and East (1997) investigated the correlation between the RAE scores and the usage of BIDS ISI databases. They found that there was a strong correlation between the two factors. There is, of course, no implication that using the ISI databases on BIDS will lead to a high RAE score. Rather, research intensive departments that are likely to score well in the RAE are also likely to be intensive users of electronic information resources.

Zhu *et al.* (1991) carried out a review of research in a few chemistry and chemical engineering departments in British universities. They found a correlation between both citation counts and publications counts, and peer review rating.

Of itself, the correlation between citation counts and RAE scores is not particularly surprising. It is well established that high citation counts reflect the impact, and by implication the importance of a particular publication, or a group of publications. It is therefore hardly surprising to find that the RAE scores, which reflect peer group assessment of the quality of research, are correlated with citation counts, which reflect the same phenomenon. What is perhaps surprising is that the correlation applies in such a wide range of subject areas, including those where citations occur relatively rarely, and ranging from those with small publications outputs to those with high ones.

The results of the three studies discussed earlier (Oppenheim, 1995, 1997; Seng & Willett, 1995) demonstrated that there is a strong correlation between citation counts and the RAE ratings of UK academic departments. However, all three studies were analyses after the event. To our knowledge, no research has ever been undertaken to predict the outcome of a forthcoming RAE by means of citation counting.

4. OUR RESEARCH

4.1 Introduction

The aim of this research was to predict *the order* of RAE ratings that will be achieved by Departments of Librarianship and Information Management in 2001, using citation analysis alone. When the RAE results are announced in 2001, comparison with the predictions made here can then be carried out.

A list was made of the members of staff in the Information and Library schools in the UK likely to be making a return under UoA 61.

The institutions examined were:

- The Robert Gordon University, Aberdeen;

- University of Wales, Aberystwyth;

- Queen's University, Belfast;

- University of Central England, Birmingham;

- University of Brighton;

- Queen Margaret College, Edinburgh;

- University of Strathclyde, Glasgow;

- Leeds Metropolitan University;

- Liverpool: John Moores University;

- City University, London;

- University of North London;

- University College, London;

- Loughborough University;

- The Manchester Metropolitan University;

- University of Northumbria at Newcastle;

- University of Sheffield.

There were a number of other institutions without library schools, departments of information science or similar titles who put in submissions to the last RAE under the heading of Library and Information Management. These were: University of Bath, de Montfort University, University of Central Lancashire, University of Leicester, and the Royal Free Hospital School of Medicine. We did not know what individuals had been put forward under these submissions, and so these institutions were not considered for this study.

It is not clear yet who will put in submissions under Library and Information Management for the 2001 RAE, but the list of institutions we examined include the most probable ones. Clearly, if many other HEIs submit under UoA 61, or if some of the Departments we examined do not, the predictions made in this paper will be weakened.

The web sites of each department were visited to obtain a list of all current academic staff. The data was collected over two days in November 1999. Professors, readers, senior lecturers, tutors, lecturers, part-time lecturers, research fellows, and heads of departments were included. Honorary professors or lecturers, visiting professors or lecturers, computer supervisors, research staff, computer technicians, secretaries and administration officers were ignored. Of course, academic staff who might be returned under the UoA by the HEI, but were not noted in the departmental list were not included. However, from our personal knowledge, research-active individuals who were Deans and Pro-Vice Chancellors and whose "home" was one of the departments checked were not omitted, even if their name was not on the departmental home page in question. For example, J. Elkin, former Head of Department at University of Central England, Birmingham, is now Dean of Faculty there, and is not on the list of staff in the Department. However, it is likely that she will be included in the RAE, and so the analysis included her.

Next, the names of staff from this list were used for *Science Citation Index* (SCI) and *Social Sciences Citation Index* (SSCI) searches on MIMAS [1] or (in just two cases - see below) on BIDS [2]. A search was carried out to see how many times each individual had been cited. Only papers they had published since 1994 were examined for citation counting purposes. MIMAS automatically de-duplicated hits from SCI and SSCI. For the two BIDS searches, we de-duplicated by inspection. When doing the citation counts, reviews of books written by the target individuals were ignored. The citation searches were carried out over a short period in January 2000.

Citations to each individual as first-named author of a paper were used. This was both because full publication lists for many of the individuals were not available, and because of the complexities of trying to share a single citation between the authors of a multi-author paper. MIMAS permits the searching for citations to an author irrespective of where his or her name appears in the order of authors in a multi-authored paper.

However, the methodology methodology described in this paper relies purely on the ranking of Departments by the RAE rating and their ranking by citation count — not the actual members involved.numbers involved. The previous studies (Seng & Willett, 1995) showed that this approach was robust.

We used BIDS only for J. Day (University of Northumbria at Newcastle) and A. Morris (Loughborough University), because of slow response time of MIMAS when handling these two very common names. As an aside, it is a matter of some concern that the new system, MIMAS, has much greater difficulty processing searches involving authors with common names than the older system, BIDS.

It should be noted that this research was carried out at the end of 1999 and in early 2000. Since then, no doubt many of the authors have received more citations, some academics have moved departments, some new staff have been recruited and others have retired.

4.2 Problems we encountered

One of the main problems with with this methodology was carrying out searches on common names. Other authors with exactly the same surnames and initials in quite different disciplines had been cited, and their citations were being duly recorded in the search. Such false drops were removed by inspection where the bibliographic details of the articles cited were clearly outside the field of librarianship or information science. However, in some many cases, there was genuine reason for doubt, as the journal or book title was ambiguous, or was in a cognate area to the one being considered. In such cases, the citations were included. There can be no doubt that some error is introduced by this, as demonstrated by one of our referees who tested our results and found that in his/her judgement, the numbers of hits were not identical to the numbers we found. However, the alternative, to obtain copies of the cited articles in doubt and to check that the the cited author's affiliation, or obtaining master lists of publications directly from each author in the library and information science departments, would have significantly added to the time and resources needed for the research. For this reason, the results presented below must be treated with caution.

On the other hand, it could be that citations for a different individual (say a US author) with the same name and initials who is active in librarianship or information management were credited to the UK academic under study. These two effects will, to some extent, cancel each other out. The errors introduced by the methodology are likely to be minor, bearing in mind that the research was not to find absolute figures, but rankings. However, as will be demonstrated below, some of the citation counts for Departments we obtained were very similar to each other. Our results, therefore, can only offer a fairly coarse differentiation between the Departments we studied.

The surname and initials as supplied by the departmental web sites were used. Hyphenated names were merged for the citation search; for example, Yates-Mercer became Yatesmercer, to follow the ISI (the producers of SCI and SSCI) conventions.

4.3 The calculations

When the citation counts had been obtained, total number of citations for each department and the average number of citations per member of staff for each department were calculated. The Departments were then ranked in order of citations, and in order of the average number of citations per member of staff.

4.4 Results

Three hundred and thirty eight staff were checked for citations. Two hundred and thirty five of these received no citations at all during the period in question. As an aside, one might wonder what the Higher Education Funding Councils would make of a Unit of Assessment where two thirds of the academics appear to have had no impact on their fellow academics at all. Twenty-six academics received one citation, twelve received two citations, seven received three citations, six received four citations and two received five citations. Fifty academics received more than five citations. In all, 2,301 citations were noted. Full details of our results, including individual scores, can be obtained from the senior author.

4.5 The effect of recent staff changes

There have been recent moves by E. Davenport and H. Hall from Queen Margaret College, Edinburgh to Napier University, of D. Ellis from Sheffield to Aberystwyth and of M. Collier (formerly in the private sector, and who was cited seven times in the period) to Northumbria University. We analysed how much an institution is vulnerable to the departure of, or benefits from, the arrival of such individuals.

Table 1 shows the rankings for numbers of citations received before these changes and Table 2 shows the rankings for numbers of citations received *after these changes.*

RANK	INSTITUTION	NUMBER OF CITATIONS	PERCENTAGE OF TOTAL (%)
1	University of Sheffield	843	36.5
2	City University, London	605	26.2
3	University of Loughborough	338	14.6
4	Queen Margaret College, Edinburgh	104	4.5
5	University of Wales, Aberystwyth	78	3.4
6	Queen's University, Belfast	62	2.7
7	University of Strathclyde, Glasgow	57	2.5
8	Northumbria at Newcastle	52	2.3
9	Manchester Metropolitan University	46	2.0
10	Leeds Metropolitan University	40	1.7

11	Robert Gordon University, Aberdeen	27	1.2
12	University of Central England, Birmingham	25	1.1
13	University College, London	11	0.5
14	University of Brighton	10	0.4
14	University of North London	10	0.4
16	Liverpool John Moores University	3	0.1
	TOTAL	**2311**	**100**

Table 2: Citations received by each department following the major staff changes

RANK	INSTITUTION	NUMBER OF CITATIONS	PERCENTAGE OF TOTAL (%)
1	University of Sheffield	726	32.1
2	City University, London	605	26.7
3	University of Loughborough	338	14.9
4	University of Wales, Aberystwyth	195	8.6
5	Queen's University, Belfast	62	2.7
6	Northumbria at Newcastle	59	2.6
7	University of Strathclyde, Glasgow	57	2.5
8	Queen Margaret College, Edinburgh	56	2.5
9	Manchester Metropolitan University	46	2.0
10	Leeds Metropolitan University	40	1.8

The departure of E. Davenport and H. Hall from Queen Margaret College, Edinburgh, to Napier University had a dramatic effect on the College, which dropped from fourth to eighth place. D. Ellis' departure from Sheffield to Aberystwyth did not affect Sheffield's top position, but helped Aberystwyth. We do not know if Napier University will return Davenport and Hall under this UoA. Collier's arrival at Northumbria has an effect, moving the University from eighth to sixth place, but this is primarily because it is in a group of departments with very similar citation totals.

A comparison of Table 1 with Table 3 shows some minor changes in order, in particular, a switch at the top between Sheffield University and City University, but there is little practical effect.

The average number of citations per member of staff taking into account the recent major staff changes is shown in Table 4.

Table 3: Average number of citations per member of staff received by each departmntt before the major changes

RANK	INSTITUTION	AVERAGE CITATIONS PER STAFF MEMBER
1	City University, London	50.42
2	University of Sheffield	49.6
3	University of Loughborough	16.9
4	University of Strathclyde, Glasgow	6.33
5	Queen Margaret College, Edinburgh	5.78
6	University of Wales, Aberystwyth	4.33
7	Queen's University, Belfast	2.48
8	Northumbria at Newcastle	1.93
9	Manchester Metropolitan University	1.92
10	University of Central England, Birmingham	1.67
11	Leeds Metropolitan University	0.98
12	Robert Gordon University, Aberdeen	0.93
13	University College, London	0.61
14	University of North London	0.4
15	Liverpool John Moores University	0.33
16	University of Brighton	0.3
	TOTAL	146

Table 4: Average number of citations per member of staff taking into account recent major staff changes

RANK	INSTITUTION	AVERAGE CITATIONS PER STAFF MEMBER
1	City University, London	50.42
2	University of Sheffield	45.38
3	University of Loughborough	16.9
4	University of Wales, Aberystwyth	10.26
5	University of Strathclyde, Glasgow	6.33
6	Queen Margaret College, Edinburgh	3.5
7	Queen's University, Belfast	2.48
8	Northumbria at Newcastle	2.11
9	Manchester Metropolitan University	1.92
11	Leeds Metropolitan University	0.98
12	Robert Gordon University, Aberdeen	0.93
13	University College, London	0.61
14	University of North London	0.4
15	Liverpool John Moores University	0.33
16	University of Brighton	0.3

A comparison of Table 3 with Table 4 shows that the only significant change in order as a result of recent staff changes was the improvement in Aberystwyth's position.

Table 5 shows the thirty most heavily cited authors identified in this study.

Table 5: The thirty most cited UK authors

RANK	NAME	INSTITUTION	NUMBER OF CITATIONS
1	S. E. Robertson	City University, London	439
2	P. Willett	University of Sheffield	410
3	D. Ellis	University of Wales, Aberystwyth	117
4	M. Hancock-Beaulieau	University of Sheffield	114
5	C. Oppenheim	University of Loughborough	102
6	T. D. Wilson	University of Sheffield	80
7	D. Bawden	City University, London	78
8	E. Davenport	Queen Margaret College, Edinburgh (since moved)	47
9	M. Lynch	University of Sheffield	46
10	G. McMurdo	Queen Margaret College, Edinburgh	41
11	A. J. Meadows	University of Loughborough	38
12	G. Philip	Queen's University, Belfast	37
13	J. Feather	University of Loughborough	37
14	E. M. Keen	University of Wales, Aberystwyth	36
15	D. Nicholas	City University, London	34
16	M. H. Heine	Northumbria at Newcastle	33
17	P. Brophy	Manchester Metropolitan University	27
18	S. Jones	City University, London	26
19	N. Ford	University of Sheffield	25
20	S. Walker	Leeds Metropolitan University	24
20	A. Morris	University of Loughborough	24
22	F. Gibb	University of Strathclyde	23
23	P. F. Burton	University of Strathclyde	22
24	C. McKnight	University of Loughborough	23
25	B. Usherwood	University of Sheffield	22

It should be noted that Professor Robertson only spends one day a week at City University but will, no doubt, be returned by City University for the 2001 RAE being considered to be on its staff list.

4.5 Predicting the RAE score order

Based upon our results taking into account the recent changes of staff, and assuming that total numbers of citations, or average numbers of citations per member of staff can be used to predict the score order, *we suggest that the order shown in Table 6 is likely to be seen in the 2001 RAE:*

Table 6: Our preditions for the rank order for the departments

1. Sheffield or City
2. Sheffield or City
3. Loughborough
4. Aberystwyth
5. Strathclyde or Belfast
6. Strathclyde or Belfast
7. Queen Margaret College or Northumbria
8. Queen Margaret College or Northumbria
9. Manchester Metropolitan
10. Leeds Metropolitan
11. University of Central England
12. Robert Gordon University
13. University College London
14. University of North London
15. University of Brighton
16. Liverpool John Moores

One can speculate regarding the actual RAE scores to be achieved by these HEIs. Twenty-three institutions were considered in this UoA in the 1996 RAE. The breakdown of RAE scores was as shown in Table 7 below:

Table 7: Breakdown of the 1996 RAE scores

1996 RAE RATING	NUMBER OF INSTITUTIONS	PERCENTAGE (%)
5*	2	8.6
5	1	4.4
4	2	8.6
3a	3	13.0
3b	7	30.4
2	5	22.0
1	3	13.0
TOTAL	23	100

We have analysed just 16 HEIs for this study. *Assuming a similar percentage breakdown* of 5*, 5, 4, 3a, 3b, 2 and 1 scores leads the following prediction (Table 8) for the distribution of RAE scores in 2001.

Table 8: A possible distribution of RAE scores in 2001

RATING	NO. OF INSTITUTIONS WITH THIS RAE SCORE>
5*	1
5	1
4	1
3a	2
3b	5
2	4
1	2
TOTAL	16

On this basis, a possible distribution of scores could be:

- Sheffield or City   5*

- Sheffield or City   5

- Loughborough   4

- Aberystwyth   3a

- Strathclyde   3a

- Belfast   3b

- Queen Margaret College   3b

- Northumbria   3b

- Manchester Metropolitan   3b

- Leeds Metropolitan   3b

- Robert Gordon   2

- University of Central England   2

- University College London   2

- University of North London   2

- University of Brighton   1

- Liverpool John Moores   1

However, it must be stressed that there is no particular reason why the 2001 distribution of scores should be similar to that for the 1996 RAE. *We emphasise that the prediction from our work is shown in Table 6.*

5. DISCUSSION

We have already noted that moves by key members of staff can have an impact of likely RAE performance. Hancock-Beaulieau (cited 114 times) recently moved from City University to Sheffield University. This increased the total number of citations for Sheffield from 612 to 726 and decreased City's from 719 to 605. The cases of Davenport and Ellis are particularly pertinent, as they are also highly cited but moved either from, or to, middle ranking HEIs. Indeed, even less heavily cited individuals can have an important impact on the middle ranking and weaker departments.

A key factor is the different citation patters for different parts of what is a very broad discipline. Robertson, Ellis and Beaulieu all work in the computer science end of LIS, which is likely to be more heavily cited than more traditional library oriented research, and this comment applies even more to work in chemoinformatics undertaken by Willett. This fact will tend to skew the figures substantially.

We suggest that some departments may be dependent upon one individual. We have calculated that S.E. Robertson of City University has received 439 citations. City University currently has a total of 605 citations and the loss of S.E. Robertson would mean that City's total would decrease to 166, and thus move down that ranking to fourth. The average citations per member of staff would decrease from 50.42 to 15.1 and so City would move down that ranking from first to third. This shows that City University would be vulnerable to the loss of this individual. Similar analyses demonstrate the dependence of Sheffield on P. Willett, Aberystwyth on D. Ellis, Belfast on G. Philip, Leeds Metropolitan on S. Walker, University of North London on R.E. Burton, Manchester Metropolitan on P. Brophy and Northumbria on M.H. Heine.

Table 5 showed the current thirty most cited UK authors in library and information science. In 1992, D. Ellis was the most cited author with 71 citations. M. Hancock-Beaulieu was second, C. Oppenheim was third and P. Willett was fourth (Oppenheim, 1995). There has been a large increase since 1992 in the number of citations gained, but the order of the top few has hardly changed. The four most cited authors in 1992 are amongst the top five most cited authors currently. It is clear that there has been little change in this regard in the past decade.

Comparing the citation counts in 1992 (Oppenheim, 1997) with our results, it is clear that Loughborough has remained in third place, but the difference between the scores for Loughborough and the top two Departments (City and Sheffield) has grown.

One can use the methodology to identify those individuals with very low citation counts. These might be considered for exclusion from the RAE return by their HEIs, though how useful such an approach would be is open to debate. On the one hand, it could ensure a higher RAE score. On the other hand, funding from the Funding Councils is based on a calculation multiplying a figure linked to the RAE rating to the number of research-active staff returned, so this means that less money may be received. There is an argument that the psychological impact of a higher RAE score both internally and externally is more important than the risks involved in returning fewer research-active staff. This is just one of many factors in RAE gamesmanship that HEIs have to consider.

Other analyses could be carried out on the actual publications selected to be returned in the RAE. At the end of the 2001 RAE, the Funding Councils will put a list of all publications submitted in the RAE on its Web site. A citation count could be carried out on just those publications, and compared to other publications authored by the same member of staff, to see if perhaps other publications may be more appropriate for the return. Such a method should be used with great caution however, as recently published papers will receive fewer publications than older ones.

Another analysis could be by Impact Factor of the journals in which the articles published by the academics have appeared. This is defined as the average number of citations received per paper by a journal. *Journal Citation Reports*, which is published by ISI, lists journals by Impact Factor within each discipline. Whilst, again, such data must be used with caution (see Amin & Mabe, 2000 for a recent critical review of the topic), as review journals get extremely high Impact Factors, yet review articles are not appropriate for return for the RAE, there is an argument that the RAE return includes a high proportion of high Impact Factor journals.

Use of citation counting could lead in future to an assessment of how departments are currently performing, and will be able to predict their performance in the RAE relative to close competitor departments. This in turn will lead to possibilities for remedial action, if considered appropriate, for weak departments.

Further work is underway for a similar study in another subject area, sports science (Iannone & Oppenheim, [unpublished]).

5.1 Potential criticisms of this work

This research is open to a wide variety of criticisms. These range from the fundamental to the detailed. For example, we became aware of the inadequacy of some of the web sites we inspected. Some were not complete and may not have been up to date. For example, E. Davenport and H. Hall were still listed at the time of our onresearch on the Queen Margaret College staff list even though they had both departed. Another problem was the effect of changing names. There were a number of authors who were cited both in maiden names and married names. For example, K. Henderson of Strathclyde, was previously K. de Placido. We knew a few such examples, and in such cases we used all the alternative names, but undoubtedly there were more that we did not recognise. Our methodology can take no account of name changes such as these.

At times, we had to make arbitrary judgements regarding whether a citation was to our chosen academic, or to someone in a different subjectarea that happened to have the same name. This remains an important weakness for any such study that does not have to hand a comprehensive master list of all publications by the academics under study.

Another criticism is that the research was undertaken some time ago, and would be better done much closer to the timing of the RAE.

A potential criticism is that this technique is usurping the peer group evaluation task. We believe that citation counts should be used to create a rough first go at the rank ordering for, but not the actual RAE scores to be assigned. This must remain a task for the panel of experts. However, we believe the use of citation counts would reduce the overall costs of the RAE, help speed up the panel's work and would reduce the amount of paperwork that the HEIs submit.

Another potential criticism is that that the exercise is too heavily tied to traditional print published materials, and that any such assessment should include analyses of Web links to Departmental Web pages as well. Thomas and Willett (2000) have described the first attempt at such an analysis.

A final potential criticism must be addressed: by publishing these results in advance of the RAE panel's decision-making, we may influence the UoA 61 RAE panel either to follow our suggested order, or to deliberately choose a different order in order to disprove our predictions. However, this criticism implies that this paper would influence the RAE panel. We believe that the panel's methods of working, and its integrity in its approach to its work, means it would not take the slightest notice of this article. In other words, we are convinced that our results should not, and will not, affect the peer judgment of the RAE panel either positively or negatively.

ACKNOWLEDGEMENTS

We would like to thank our two anonymous referees for their helpful comments.

NOTES

1. MIMAS. (Manchester Information & Associated Services) "MIMAS is a JISC-supported national data centre providing the UK higher education, further education and research community with networked access to key data and information resources to support teaching, learning and research across a wide range of disciplines."

2. BIDS. (Bath Information and Data Services) "BIDS is the best known and most widely used bibliographic service for the academic community in the UK. BIDS also provides access to the ingentaJournals full text service, both directly and also through links from database search results."

REFERENCES

1. Ajibade, B. & East, H. (1997) *Research Assessment Exercise and usage of BIDS ISI.* London: The City University, Database Resources Research Group.

2. Amin, M. & Mabe, M. (2000) Impact factors: use and abuse, *Perspectives in Publishing*, (No. 1) (Entire issue)

3. Baird, L.M. & Oppenheim, C. (1994) Do citations matter? *Journal of Information Science*, 20, 2-15.

4. Carpenter, M.P., Gibb, F., Harris, M., Irvine, J., Martin, J.R. & Narin, F. (1988) Bibliometric profiles for British academic institutions: an experiment to develop research output indicators. *Scientometrics*, 14(3/4), 213-233.

5. Cole, J.R. & Cole, S. (1973) *Social stratification in science.* Chicago, London: University of Chicago Press.

6. Colman, A.M., Dhillon, D. & Coulthard, B. (1995) A bibliometric evaluation of the research performance of British university Politics departments: publications inleading journals. *Scientometrics*, 32 (1), 49-66.

7. Garfield, E. (1979) *Citation indexing: its theory and application in science, technology, and humanities.*New York: Wiley Interscience.

8. HEFCE (1994) *Circular 1/94: 1996 Research Assessment Exercise.* Bristol.: HEFCE, SHEFC, HEFCW, DENI.

9. HEFCE (1995) *1996 Research Assessment Exercise: guidance on submissions*Bristol: HEFCE.   [http://www.niss.ac.uk/education/hefc/rae96/c2_95.html] [Accessed 22/12/2000]

10. Hemlin, S. (1996) Research on research evaluation. *Social Epistemology*, 10(2), 209-250.

11. Iannone, R. & Oppenheim, C. [Unpublished results]

12. Liu, M. (1994) The complexities of citation practice: a review of citation studies. *Journal of Documentation*, 49, 370-408.

13. Nederhof, A.J. & Noyns, E.C.M. (1992) Assessment of the international standing of University Departments' research: a comparison of bibliometric methods. *Scientometrics*, 24(3), 292-304.

14. Oppenheim, C. (1979) Could the 1978 Nobel prizewinner in chemistry have been predicted?, *Social Studies in Science*, 9, 507-508.

15. Oppenheim, C. (1995) The correlation between citation counts and the 1992 Research Assessment Exercise Ratings for British library and information science University Departments. *Journal of Documentation*, 51(1), 18-27.

16. Oppenheim, C. (1997) The correlation between citation counts and the 1992 Research Assessment Exercise ratings for British research in Genetics, Anatomy and Archaeology. *Journal of Documentation*, 53(5), 477-487.

17. Seng, L.B. & Willett, P. (1995) The citedness of publications by United Kingdom library schools. *Journal of Information Science*, 21, 68-71.

18. Summers, R., Oppenheim, C., Meadows, J., McKnight, C. & Kinnell, M. Information science in 2010: a Loughborough University view. *Journal of the American Society for Information Science*, 50 (12), 1153 — 1162.

19. Thomas, O. & Willett, P. (2000) Webometric analysis of departments of librarianship and information science, *Journal of Information Science*, 26 (6), 421-428.

20. Winclawska, B.A. (1996) Polish sociology citation index: principles for creation and the first results. *Scientometrics*, 35(3), 387-391.

21. Zhu, J., Meadows, A.J. & Mason, G. (1991) Citations and departmental research ratings. *Scientometrics*, 21, 171-179.

CHAPTER 14

A Full Text Retrieval System in a Digital Library Environment

Kehinde Daniel Aruleba[1], Dipo Theophilus Akomolafe[2*], Babajide Afeni[3]

[1]*Department of Mathematics & Computer Science, Elizade University, Ilara-Mokin, Nigeria*

[2]*Department of Mathematical Sciences, Ondo State University of Science and Technology, Okitipupa, Nigeria*

[3]*Department of Computer Science, Joseph Ayo Babalola University, Ikeji-Arakeji, Nigeria*

ABSTRACT

The volume of information being created, generated and stored is huge. Without adequate knowledge of Information Retrieval (IR) methods, the retrieval process for information would be cumbersome and frustrating. Studies have further revealed that IR methods are essential in information centres (for example, Digital Library environment) for storage and retrieval of information. Therefore, with more than one billion people accessing the Internet, and millions of queries being issued on a daily basis, modern Web search engines are facing a problem of daunting scale. The main problem associated with the existing search engines is how to avoid irrelevant information retrieval and to retrieve the relevant ones. In this study, the existing system of library retrieval was studied. Problems associated with them were analyzed in order to address this problem. The concept of existing information retrieval models was studied, and the knowledge gained was used to design a digital library information retrieval system. It was successfully implemented using a real life data. The need for a continuous evaluation of the IR methods for effective and efficient full text retrieval system was recommended.

KEYWORDS

Full text, Information Retrieval, Library, Digital Library, Queries, Indexing, Catalogue

1. INTRODUCTION

For Centuries, libraries have been organizing reading materials on shelves for easy access. However, systematic methods that had been widely adopted for the organization of library materials and their recordings for use by readers came into being a little more than a century ago [1] . The term digital library is used to refer to a library where some or all of the holdings are available in electronic form, and the services of the library are also made available electronically-frequently over the Internet so that users can access them remotely [2] . The primary purpose of digital libraries is to enable searching of electronic collections distributed across networks, rather than merely creating electronic repositories from digitized physical materials.

An information retrieval (IR) system is designed to retrieve any documents or information required by the user community. It is primarily targeted to make the right information available to the right user at right time. IR is concerned with representing, searching, and manipulating large collections of electronic text data. IR is a discipline that deals with retrieval of unstructured data or partially structured data, especially textual documents, in response to a set of query or topic statement(s), which may itself be unstructured [3] . IR system does not inform i.e. change the knowledge of the user on the subject of his enquiry; it merely informs the user of the existence or non-existence and whereabouts of documents relating to the request.

Many problems are associated with the current system of IR and such can be seen from the inability of the system to process request timely and to present inadequate results among others. In view of these inadequacies, it is imperative to develop an IR system that will curtail these inadequacies.

2. RELATED WORK

The importance of IR keeps growing as the amount of digital information keeps expanding at an ever-increasing rate. Stored documents, photographs and contents of books, and billions of Web pages are useful only if they can be easily found when needed.

2.1. Information Retrieval Models

For effectively retrieving relevant documents by IR strategies, the documents are typically transformed into a suitable representation. Each retrieval strategy incorporates a specific model for its document representation purposes. According to [4] , the Boolean model is the first model of IR and probably also the most criticized model. Larson [5] shows that much of this criticism seems to be based on lack of knowledge about how to utilise its search possibilities. In this model, we can pose any query which is in the form of a Boolean expression of terms, that is, in which terms are combined with the operators AND, OR, and NOT. The model views each document as just a set of words.

The vector space model (VSM) represents documents and queries as vectors in multidimensional space, whose dimensions are the terms used to build an index to represent the documents. It is used in IR, indexing and relevancy rankings and can be used in evaluation of Web search engines. According to Shang [6] , the VSM procedure was divided into three stages. The first stage is the document indexing where content bearing terms are extracted from the document text. The second stage is the weighting of the indexed terms to enhance retrieval of document relevant to the user. The last stage ranks the document with respect to the query according to a similarity measure.

According to Gonzalez [7] , Language models (LM) for information retrieval are retrieval models (taken from the speech recognition field) that do not impose an explicit parametric form for the probability of relevance. Lafferty and Zhai [8] presented a formal connection between probabilistic and language models. The basic idea of the language modelling approach to IR is to assume that a query Q is generated by a probabilistic model of document D. In this context, the generative language models approach estimate $£_i$ is the probability of the query being generated by a document.

2.2. Query Types

There are many different ways of searching for information. Here we describe the most common ones according to Salerma [9] .

A normal query is any query that is not explicitly indicated by the user to be a specialized query. For queries containing only a single term, the desired semantics are clear: match all documents that contain the term. For multi-word queries, however, the desired semantics are not so clear.

Some implementations treat it as an implicit Boolean query by inserting hidden AND operators between each search term.

Phrase queries are used to find documents that contain the given words in the given order. Usually phrase search is indicated by surrounding the sentence fragment in quotes in the query string. They are most useful for finding documents with common words used in a very specific way [9] . For example, if you do not remember the author of a paper, searching for it on the Internet as a phrase query will in all likelihood find it for you.

Boolean queries are queries where the search terms are connected to each other using the various operators available in Boolean logic, most common ones are AND, OR and NOT [10] . Usually parentheses can be used to group search terms. A simple example is software AND database, and a more complex one is software AND database AND data structure.

3. INFORMATION RETRIEVAL AND DIGITAL LIBRARIES

Libraries have been in existence since the beginning of writing and have served as a repository of the intellectual wealth of society. As such, libraries have always been concerned with storing and retrieving information in the media it is created on. As the quantities of information grew exponentially, libraries were forced to make maximum use of IR methods to facilitate the storage and retrieval process. Some of the IR methods used in digital libraries are described in the following:

Indexing: IR systems need an indexing mechanism for performing efficiently the retrieval process [7] . Indexing is the transformation from the received item to the searchable data structure. Building an index from a document collection involves several steps, from gathering and identifying the actual documents to generating the final data structures [11] .

Catalogue records: are short records that provide summary information about a library object. The word catalogue is applied to records that have a consistent structure, organized according to systematic rules. Library catalogues serve many functions, not only information retrieval. Some catalogues provide comprehensive bibliographic information that cannot be derived directly from the objects. This includes information about authors or the provenance of museum artefacts [12] . Descriptive Metadata: Many methods of information discovery do not search the actual objects in the collections, but work from descriptive metadata about the objects [12] . The metadata typically consists of a catalogue or indexing record, or an abstract, one record for each object. Usually it is stored separately from the objects that it describes, but sometimes it is embedded in the objects. Descriptive metadata is usually expressed as text, but can be used to describe information that is in formats other than text, such as images, sound recording, maps, computer programs, and other non-text materials, as well as for textual documents.

3.1 Library Digitization

According to Ian and David [13] , defined digitization as the process of taking traditional library materials that are in form of books and converting them to the electronic form where they can be stored and manipulated by a computer.

According to Alhaji [14] , there are three main reasons for digitization of a library system

1   To make the documents more accessible: This is to serve existing library users better, i.e. to allow users search the full text of documents or to allow users search from remote locations.

2 To preserve the documents: Allow user read older or unique documents without damage to the originals

3 To reuse the documents: Allowing conversion of documents into different formats.

4. METHODOLOGY

From the architecture of a FTRS described in Aruleba et al. [15] , a full text search retrieval system was designed. This section presents the modelling of the system. The modelling is in two parts, which are: Analysis and Design

The analysis of the existing information system in University of Ilorin library was extensively carried out by studying the existing environment. The result of the analysis is presented using the Use-case diagram shown in figure 1.

The result of the existing library information shows that the entire system is made of seven steps. The first step is where the potential library user is registered. The registration allows the user to become a registered and legally authorised user of the library. After the registration, the user is allowed to use the facilities provided by the library. After registration, the registered user is allowed to undertake the remaining steps that is registered user can check available books, read books, check document title, author, publisher, borrow book, return book and pay fine in case of late submission of book(s).

The proposed system in addition to the functionality of the existing system allows users to search, modify user details, and upload documents as shown with use-case in Figure 2.

From Figure 2, the proposed system is made up of eight distinct steps. Though the components are interwoven, each of them performs distinct functions but all work together as a system to process request timely.

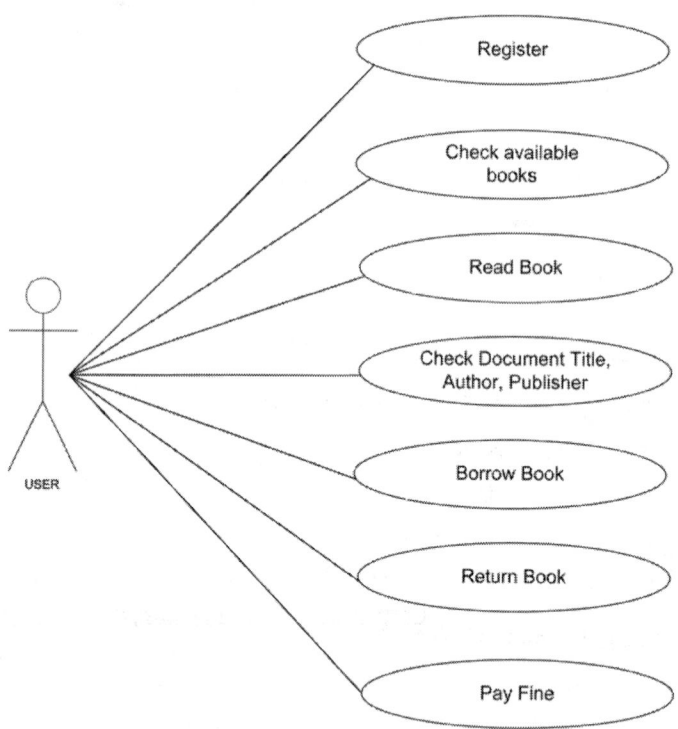

Figure 1. The result of analysis of UNILORIN library system.

4.1. Database Design

Database design mainly includes requirement analysis, concept structure design stage, the logic structure design stage, physical structure design stage, database implementation stage, database operation and maintenance stage, there are six steps altogether.

From the analysis done, Table 1 was designed for the implementation of the proposed system.

4.2. User Interface Design

There are many factors that must be considered when designing the user interface of a software because the user must be able to interact with the system in a way that the system will understand whatever input given by the user. Therefore, the quality of the interface and software in general must pass the usability testing standard. Some usability factors, such as fit for use, ease of learning, task efficiency, ease to remember, subjective satisfaction and understand ability but all are put into consideration when designing the user interface (Figure 3).

The home page screen depicted in figure 4, contains four major modules which are the Search, Registration, Request and Login while the Admin module home page shown in figure 5, contains Sub-module which are view students, view staff, view books, create new book, view book request, create/view facilities, create/view department, logout. Each of them will lead you to its database when clicked and manipulated.

5. SYSTEM IMPLEMENTATION PHASE AND TESTING

This phase implements what have been discussed in thesection 4. The system was developed and implemented with PHP and MySql Technology.

Figure 2. Use case diagram showing the proposed system.

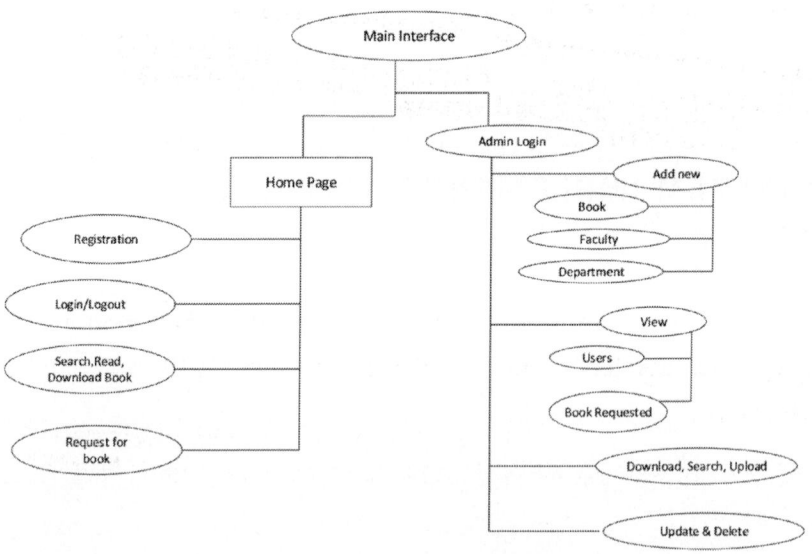

Figure 3. Showing the main interfaces of the system.

Table 1. Generated database.

Table	Action
Admin	To authenticate the system user
Book request	To store information about requested books by the system users
Books	To store book details
Department	To store all the departments available
Faculty	To store all the faculties available
Staff	To store information of staffs using the system
Students	To store information of students using the system.

Figure 4. Home page design interface.

Home Page Interface Implementation

The home page shown in figure 6 is the key aspect of the system, because it gives the basic user interface for the full text retrieval digital library. It comprises of: Search, Login, Registration and Request described as follows:

Search: This feature can be used by any user. This module provides a convenient book searching function, the user could search books based on a variety of conditions.

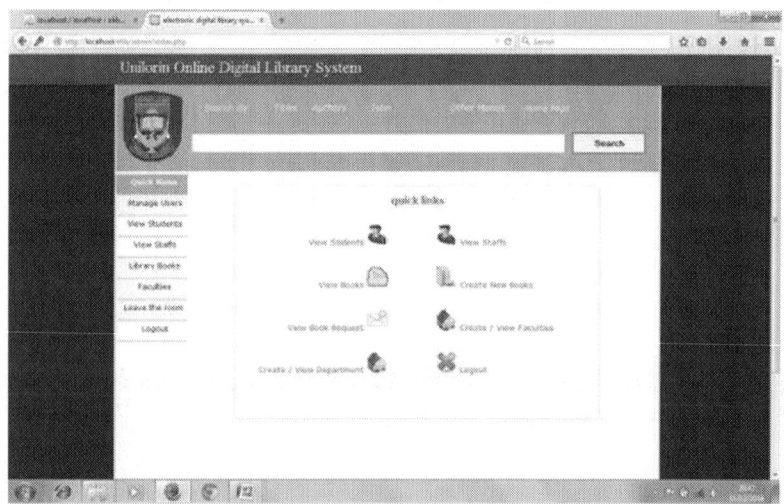

Figure 5. Admin home page design interface.

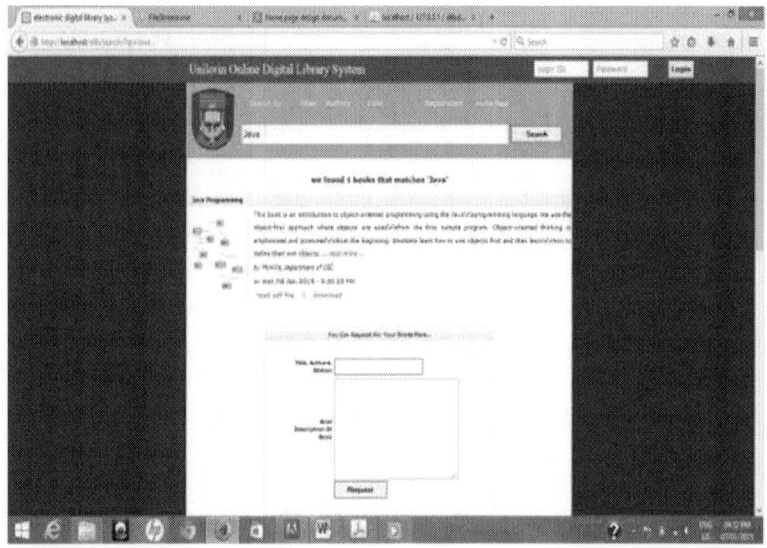

Figure 6. Home page implementation output.

Login: Every user who wants to use the system is authenticated by means of username and password. All entered parameters of the password are matched with information stored in the database, therefore only authenticated users can log on to the program with limited access.

If the login information is wrong, the user will be notified of login failure and would need to try again.

Registration: This involves registering new users. It contains registration form interface with entries like email address, last name, first name, password, password confirmation and sex.

Request: If a user can find the specific book needed, request can be made for such book.

6. CONCLUSIONS AND FUTURE WORK

This study has been able to develop and successfully implement an IR system that reduces the hurdles associated with present system of searching in Libraries. It is shown in the system that users searching full text are more likely to find relevant articles than searching only abstracts. This finding affirms the value of full text collections for text retrieval and provides a platform for aligning searching algorithms that take advantage of rapidly-growing digital archives.

Also, the following areas of the study can be improved upon in future studies to create a more robust IR:

- Increase in the size of the database, this will enable large data storage.

- Integrate advert plans for research materials the institution wants to be selling online.

- Acquire and publish video, audio and heavy graphic research materials.

REFERENCES

1. Onwuchekwa, E.O. and Jegede, O.R. (2011) Information Retrieval Methods in Libraries and Information Centers. An International Multidisciplinary Journal, 5, Serial No. 23

2. Rosenberg, D. (2005) Towards the Digital Library: Findings of an Investigation to Establish the Current Status of University Libraries in Africa. International Network for the Availability of Scientific Publications, Oxford.

3. Greengrass, E. (2002) Information Retrieval: A Survey by Ed Greengrass. Information Retrieval, 141-163.

4. Goker, A. and Davies, J. (2009) Information Retrieval: Searching in the 21st Century. Wiley, West Sussex.

5. Larson, R.R. (2010) Information Retrieval Systems. In: Bates, M.J. and Maack, M.N., Eds., Encyclopedia of Library and Information Sciences, 3rd Edition, CRC Press, New York, IV, 2553-2563.

6. Li, L.Z. and Shang, Y. (2000) A New Statistical Method for Performance Evaluation of Search Engines. ICTAI.

7. González, R.B. (2008) Index Compression for Information Retrieval Systems. Unpublished PhD Thesis, University of A Coruna, A Coruna.

8. Lafferty, J. and Zhai, C. (2003) Probabilistic Relevance Models Based on Document and Query Generation. In: Croft, W.B. and Lafferty, J., Eds., Language Modelling for Information Retrieval, Kluwer, Pittsburgh, 11-56.

9. Salerma, O. (2006) Design of a Full Text Search Index for a Database Management System. MSc Dissertation, University of Helsinki, Helsinki.

10. Pazer, J.W. (2013) The Importance of the Boolean Search Query in Social Media Monitoring Tools. Dragon Search.

11. Manning, C.D., Raghavan, P. and Schutze, H. (2008) Introduction to Information Retrieval. Cambridge University Press, Cambridge.

12. Arms, W. (2002) Manuscript of Digital Libraries. MIT Press, Cambridge.

13. Written, L.H. and Brainbridge, D. (2003) How to Build a Digital Library. Morgan Kaufman Publishers, London.

14. Alhaji, I. (2005) Digitization of Library Resources and the Formation of Digital Libraries: A Practical Approach. University of Pretoria, Pretoria.

15. Aruleba, K.D., Aremu, D.R., Oriogun, P.K., Agbele, K.K. and Agho, A.O. (2015) Evaluation of Full Text Search Retrieval System. Nigeria Computer Society, 26, 154-159.

INDEX